Design and Evaluation of Ad Hoc Routing Protocol

SCIENCES

Networks and Communications, Field Director – Guy Pujolle

Vehicular Networks, Subject Head – Selma Boumerdassi

Design and Evaluation of Ad Hoc Routing Protocol

Martine Wahl
Patrick Sondi

iSTE

WILEY

First published 2024 in Great Britain and the United States by ISTE Ltd and John Wiley & Sons, Inc.

ISTE Ltd
27-37 St George's Road
London SW19 4EU
UK

www.iste.co.uk

John Wiley & Sons, Inc.
111 River Street
Hoboken, NJ 07030
USA

www.wiley.com

Library of Congress Control Number: 2024934516

British Library Cataloguing-in-Publication Data
A CIP record for this book is available from the British Library
ISBN 978-1-78945-112-2

ERC code:
PE6 Computer Science and Informatics
PE6_2 Computer systems, parallel/distributed systems, sensor networks, embedded systems, cyber-physical systems
PE7 Systems and Communication Engineering
PE7_8 Networks (communication networks, sensor networks, networks of robots, etc.)

Contents

Preface

Intent, abstract and target audience for this book

This book provides a summary of the research work and research frameworks collaboratively carried out by the authors, namely, Martine Wahl, research fellow at the Université Gustave Eiffel and Patrick Sondi, formerly lecturer at the Université du Littoral Côte d'Opale, currently professor at the Institut Mines-Télécom Nord Europe.

The authors' intention is to share their ideas and knowledge accumulated over time, for almost 10 years, in the field of communication protocol engineering, especially for land transport systems.

This book focuses on ad hoc communications between vehicles in a road environment. The development of driver assistance applications inevitably creates a dependency in users who, in the long run, will hardly be able to travel without them. To prove this point, a vehicle without power steering, without an anti-lock braking system (ABS) and without a navigation system should be entrusted to a driver who has recently obtained their driving license to travel to a destination located more than 200 km away through different types of roads. This book raises the issue of the continuity of communication-dependent driving assistance services in the road framework in areas where no communication infrastructure has been deployed or has become inoperative. With the help of an example of an ad hoc routing protocol that the authors helped design, this book presents the methodology that was followed from its design to its evaluation. The authors outline the questions and motivations involved in their functional requirement-based design approach to be met through the protocol. A description of the analyses carried out to understand the operation of the protocol, its properties as well as its performance for target applications is proposed in the various chapters of the book. The latter concludes with the prospect of integrating components managed by a protocol based on ad hoc communications into a global telecommunications infrastructure for the road.

It is aimed in particular at students and researchers who are beginning in the fields of protocol engineering, ad hoc networks or intelligent transport systems. Nevertheless, it can provide specialists in the field with another point of view on the scientific literature related to these fields, or even concrete tools for developing their own design and evaluation methods of communication protocols for different networks.

Book organization

The book was written by the two authors based on summaries of scientific studies, activity reports, translations of scientific articles, all reorganized and often rewritten in order to ensure coherence in the work. It consists of approximately 250 pages organized into six chapters.

Chapter 1, entitled "Challenges Involving Ad Hoc Communications on the Road", can be seen as a general introduction to the book. The idea is to briefly expose the evolution of road communications by addressing all aspects: traffic, protocols, applications and security.

Chapter 2, entitled "Functional Approach to Protocol Design", describes the functional approach to protocol design based on a protocol proposed as a case study for the purposes of analyzing and illustrating the design methods from the functional and performance evaluation requirements followed in this approach. The design and evaluation steps of the case study protocol are described and justified step by step, accompanied by resources in the literature so that the reader's documentation can be broadened.

Chapter 3, entitled "Models and Simulations for Performance Analysis", presents a series of models and simulation tools for protocol performance analysis. Most of the factors affecting communications such as mobility, propagation phenomena, communication technologies as well as the metrics employed for performance analysis are explained. The corresponding models as well as their implementation in the various simulation tools are illustrated in as much detail as possible.

Chapter 4, entitled "Performance Evaluation of the Protocol for Collaborative Applications", provides several sets of evaluations that include the case study protocol alone to measure its intrinsic performance, and then by comparison with other routing protocols in the literature representing the main approaches adopted in ad hoc networks. A second set of evaluations provides a performance analysis of the case study protocol for a variety of applications, in particular for two cooperative applications, namely, ego-localization and extended perception. In addition to the description of the various protocols and applications, the reader will find the steps involved in their implementation in the simulators as well as the results of the comparisons made.

Chapter 5, entitled "Formal Model for the Analysis of the Properties of the Protocol", proposes an alternative approach based on formal modeling for designing and evaluating the protocol, no longer essentially from the performance point of view, but rather on its intrinsic properties, in particular with the aim of showing the guarantees it can provide under well-identified assumptions. The analysis of the case study protocol is proposed with the Event-B method and each modeling, the verification and animation steps of the model are illustrated by extracts of the model from the tool that was used, namely, Rodin in these cases.

Chapter 6, entitled "Integration of Ad Hoc Components in a Global Infrastructure", lists the perspectives that our work has opened by federating them around the idea of a shared infrastructure of communications networks for the future land transport system that would integrate road and rail. The authors highlight in particular the integration of vehicle-to-vehicle (V2V) communications as a native communication mode of the future technology, in this case 5G, and analyze the consequences of this major evolution on the design and evaluation methods of ad hoc communication protocols in the near future.

April 2024

1

Challenges Involving Ad Hoc Communications on the Road

1.1. Introduction

With the arrival of two key innovations that will revolutionize communications in future transportation systems, we felt it was the perfect time to present a summary of our research work, which for the most part has been collaboratively achieved over nearly a decade.

The first innovation is the fully autonomous vehicle equipped with a hyperconnected on-board computer and whose operation on the roads is scheduled for the near future. Initiated in the 1970s for guided public transport, in particular with the VAL metro in the region Hauts-de-France in France, the use of communications and autonomous vehicles has gradually expanded and increased in land transport. For example, in the 2000s, mobile communications (*Global System for Mobile communications-Railway* (GSM-R)) were introduced into the European Rail Traffic Management System (ERTMS), which made it possible to harmonize and streamline high-speed train traffic across Europe. Advances in automation and localization have enabled the Société nationale des chemins de fer français (SNCF) to forecast the autonomous freight train for the end of 2021 and to effectively start testing on rails in early 2022, including the autonomous regional passenger train. In addition, driver-assistance applications initially, offered to drivers through unconnected devices (navigation, hazard warning, etc.), have evolved into connected versions available on smartphones before being fully integrated into the now connected on-board computers of vehicles, including their cooperative versions. In parallel with this increasing connectivity, improvements in sensing systems and autonomous driving systems have made it possible to improve driving automation to the levels currently observed in some vehicle models. Following this perspective of full driving automation, safety considerations have been increased and, consequently,

Design and Evaluation of Ad Hoc Routing Protocol,

must be inevitably applied to the communication system that plays a crucial role in the operation of many of these applications for the vehicular context. Communication systems for the transport of the future will therefore need to clearly integrate both the autonomous vehicle and the ability to support all essential functionalities for the remote control of a vehicle, including that initially driven by a human (as for example in a distressful situation).

The second innovation is 5G technology, whose deployment in France is already effective at the time of writing. It completes the convergence of communications, in particular by integrating the Internet of Things (IoT) into the long list of devices already brought together by 4G through all-IP. In terms of access, 5G assimilates all modes of communication, especially those concerning vehicles, namely vehicle-to-infrastructure (V2I) and vehicle-to-vehicle (V2V), while providing very low latency and many other performances at levels largely adapted to vehicular communications. In terms of the organization of access and the core of the network, 5G incorporates most of the recent innovations, resulting in particular from the diversification of access (frequency extension and pooling), the softwarization of radio and network operations (*Software-Defined Radio* and *Software-Defined Network,* respectively) and the virtualization of network-related functions (*Network Function Virtualization*) through transferring most of the transfer operations of specific hardware devices to virtual machines, especially in the cloud. Combined with paradigms such as *edge computing*, *fog computing* and *named data networking* (NDN), 5G should enable much faster access to data as close as possible to its use, which represents a clear advantage for vehicular applications, most of which require very high responsiveness. This very wide variety of applications presuppose that 5G could soon become a mobile communication system for all smart city applications, and consequently for communications in land transport systems. Therefore, the communication systems for the transport of the future need to be rethought by taking into account the drastic changes that 5G will bring and whose major impact on the design and evaluation of these systems is foreseeable.

1.2. The railway experience in the vehicular networks of the future

In rail transport, most particularly, the question of the future communication system arises more urgently due to the obsolescence and the forecast end of the maintenance of GSM-R technology. The adoption of GSM-R as a telecommunications subsystem of the ERTMS had been validated following several years of measurement campaigns and on-site tests that could not be replicated today for cost reasons. Several research projects have proposed solutions for performing simulation-based evaluations of alternative communications technologies to GSM-R, taking into account elements of the railway context likely to affect communications (tracks, train movements, interchanged traffic, etc.) (Sondi et al. 2017). Similar solutions have been used to evaluate other communication technologies as potential candidates to replace GSM-R, including Wimax (Pinedo et al. 2015), LTE (Sniady and Soler 2014) and 5G (Ai et al. 2020).

Rail infrastructure provides the first model for large-scale deployment, compared to the underground, which often finds itself limited to the city scale, of a communication system that allows the use of fully autonomous vehicles in land transport in Europe in the short term. The harmonization already undertaken at the European level between the various national rail traffic management systems reinforces the idea that rail can be the ideal support model for the development of communication systems for all future land transport in Europe. It is a foundation that must nevertheless be observed through its successes, but also through its pitfalls, while considering its specificities to avoid biases in the conclusions to be drawn for future systems intended for all land transport. Just to mention a few of the most important aspects, we should consider the following:

– *The dedicated nature of the railway infrastructure* which currently circumscribes both the scope of investments, the investor and the beneficiary. It is a different matter when it comes to a future global infrastructure for all land transport. For example, even if it could have happened, financing the deployment of GSM-R base stations in a white zone today at a loss is already an issue if there is a safety issue for a few trains crossing it. It is not clear that in the future operators would deploy a similar infrastructure for motorists on roads in sparsely populated areas where it would not otherwise be profitable. The design of a comprehensive infrastructure will need to address this aspect.

– *State hegemony over a railway operator that facilitates local decisions within each state and consequently harmonization at the European level.* Each national historical operator actually generally imposes its technological choices at the level of its state. In addition, since the telecommunications infrastructure is a dedicated infrastructure, it also remains subject to management under the sole authority of the railway operator. It is thus easier to reach a consensus at European level with one decision-maker per state (the railway operator) rather than in a context where several actors with divergent interests have been involved. The challenge will have to be addressed differently for a global infrastructure including roads for cars, as it will involve a myriad of national land transport operators (rail, road, etc.), not to mention telecommunications operators also concerned by any infrastructure deployment in the public domain. Additionally, the latter can also comprise numerous service providers and a large variety of clients, ranging from government services to the general public, including companies of the transport sector, of which car manufacturers. This multiplicity of decision-making partners, combined with the pressure of the very varied expectations of the different beneficiaries with heterogeneous profiles, presupposes that the process of harmonizing technological and normative choices will be far more complex in the context of a global infrastructure for the land transport of the future.

– *A relative de facto insensitivity to intermediate technological developments.* Indeed, the colossal investments required to equip railway infrastructure on a European scale mean that, on the one hand, the technologies chosen have to be sufficiently mature to guarantee their sustainability, and, on the other hand, that they be need to be maintained for as long as possible over time to amortize costs. For example, it is interesting to observe that GSM-R is still being deployed on railway lines, although the end of the maintenance of the GSM technology has already been announced by manufacturers. Meanwhile, the general public has shifted through 3G, then 4G and now 5G. It will probably be necessary to wait for the generalization of this latest disruptive technology for the rail sector to start new investments in Europe to achieve a technological leap, while the respective implementation of 3G and 4G has been systematically accompanied by new applications for drivers through smartphones and on-board vehicle computers. A future global infrastructure for vehicular communications is likely to be more sensitive to technological developments. It will therefore have to be designed to allow for faster and less costly adaptation compared to what has been observed in the rail context.

1.3. The contribution of ad hoc communications to vehicular networks

Unlike rail, the problem arises differently for vehicular communications within the road environment. The lack of dedicated communication infrastructure already deployed on a large scale, similarly to GSM-R in the rail sector, leaves all options still open. Admittedly, the coverage of most urban areas and the surrounding area by communication networks open to the public ensures that vehicles have access and performance adapted to their traffic. The even greater performance, especially in terms of speed and latency, advertised with 5G as well as the extension of coverage that the latter will provide outside cities, since it will also cover agricultural and industrial areas, suggests that vehicles will be able to benefit from access to these infrastructures on more sections of the road network. Nevertheless, long stretches of uncovered roads will always exist, especially in sections located far from cities and human activity, unless dedicated infrastructure is deployed. However, we have already pointed out the difficulty for reasons of economic profitability and because of the absence of a central operator capable of assuming these portions for strategic reasons (safety, transport network efficiency), as has happened in the rail sector.

This situation implies the need to integrate into the coverage strategy of any future communication system, considered for land transport, solutions making it possible to extend the coverage and continuity of services provided to vehicles in the areas which lack them.

First, there is the possibility of moving toward shared infrastructure with other land transport networks, especially in areas where roads without public coverage cross the coverage areas of the infrastructure dedicated to rail communications. This strategy could become a real possibility because the shared infrastructure would reduce costs for the rail operator without necessarily compromising safety and security, as the new network infrastructures (4G/5G) can ensure that the complete separation of the traffic of railway applications with regard to other traffic is guaranteed.

Second, ad hoc V2V communications can be resorted to and organized in such a way that most services, including cooperative applications that only need the participating vehicles, can continue to operate even if there is no connection to centralized servers due to lack of network access. The mobile ad hoc network (MANET) paradigm has been explored since the 1990s and its first extensions to the vehicular environment (*vehicular ad hoc networks* (VANET)) emerged in the 2000s. Many approaches have been proposed, namely, involving adapted communication technologies and specific routing protocols, applications and architectures. A more detailed state of the art on these different aspects will be proposed throughout the various chapters of this book, the idea here being to highlight the issues that explain both the interest and the difficulties related to the deployment of this solution on a large scale when a large part of the scientific community has been addressing it for almost a quarter of a century.

1.3.1. *Maintaining connectivity and service continuity*

The fundamental interest of ad hoc communications is that it allows communications between vehicles to be possible where the only alternative left by the lack of infrastructure is that there are none at all. Therefore, when two arbitrary vehicles are within communication range of each other, they can detect each other and exchange information.

It is often argued that such communication, only possible in direct connection and within the scope of communication, can be regarded as anecdotal at best. When two vehicles are separated by a distance greater than the communication range offered by current wireless technologies, a distance greater than three to ten times the safety distances, currently defined by road traffic rules in most European countries, does in fact separate them. Consequently, the fact that they cannot communicate beyond that distance does not affect the exchanges necessary for avoiding immediate danger between them. As a result, all emergency notification services (emergency braking, roadway obstacle, blocked lane, slowdown, etc.) can be maintained through these ad hoc communications.

Most vehicle services are offered by providers who operate from servers, and therefore require access to a network infrastructure. In this respect, these applications will follow the evolution of applications provided in other areas, that is, they will evolve toward versions capable of supporting operation in non-connected mode for a significant period of time and of being updated in a timely manner when they are reconnected to the network. These services are generally less often necessary for safety, but rather used for practical needs (navigation, traffic event notification, traffic management, advertising, etc.). It is therefore the responsibility of each service provider to evaluate the opportunity to develop this slightly connected version capable of operating in degraded mode. Users and public bodies in the automotive world are the only ones who can create a balance of power to encourage them to do so.

1.3.2. *Cooperation and performance improvement*

Several services are increasingly being developed on a cooperative basis. For example, a current widely-used application retrieves notifications generated by all participating users at the level of the publisher's servers, which determine their relevance and plausibility, and then notifies all users located in the area of interest. It is clear that such an application could proceed between vehicles operating in the area of interest, including outside the coverage of any network infrastructure, if the publisher provided for a degraded mode of operation in which the verification of consistency and plausibility could be carried out in a distributed manner directly by the vehicles involved in exchanging through ad hoc communications.

Other so-called cooperative applications, two in particular which will be presented in detail later in this book, have been designed to operate essentially and precisely on this ad hoc communication mode between vehicles. Unlike the emergency alerting services discussed earlier (section 1.3.1), which are essentially based on single-hop operation, in a V2V fashion, cooperative applications may require the organization of multiple vehicles into disjointed groups or the relay of some V2V messages over multiple hops. Under these conditions, it is no longer simply a question of relying on ad hoc V2V communications, but of detecting possible groups of vehicles and organizing them in such a way as to allow these forms of exchange in cooperation. This task is typically accomplished by routing protocols as part of a feature referred to as "self-organization". Chapter 2 of this book will revisit this aspect in great detail.

On the one hand, the distributed nature of this self-organization requires the vehicles participating in the same group to all be on the same communication channel, at least for *broadcast* exchanges, particularly those that allow them to detect each other. Advanced mechanisms make it possible to organize communications on different channels for certain vehicles, but overall, this operation, which results in an increasing number of vehicles on the same communication channel, is sometimes the cause of performance problems. This observation is often mentioned in criticisms leveled against the use of ad hoc communications in various applications, especially for vehicular networks. Any mechanism that would lead to an improvement in this performance would therefore be beneficial to the functioning of cooperative applications. A large body of work has been carried out on more efficient self-organization, quality of service (QoS) or quality of experience (QoE) in mobile ad hoc networks in general, and vehicle ad hoc networks in particular contribute to progress in this direction.

On the other hand, when a set of vehicles is self-organized outside of the coverage area of any infrastructure and these vehicles are able to exchange efficiently in cooperative applications, the question is: what should happen to this group when it integrates an area covered by a communications network infrastructure? Actually, if the group restricts its exchanges within the framework of the structure resulting from self-organization, all associated traffic will not be addressed by the network infrastructure, which will thus keep its capacities for processing further traffic. By generalizing this observation, it can be intuitively observed that maintaining self-organized groups that correctly operate via ad hoc communications, including in areas covered by a network infrastructure, can improve the overall performance. This naturally assumes that the exchanges of the self-organized group do not take place on channels used by the infrastructure for other services, which would cause interference that could degrade part of the performance. This problem can be solved by the ability of the network infrastructure to detect channels used by self-organized groups and avoid assigning them to other exchanges. The integration of self-organized groups via ad hoc communications into a broader framework including network infrastructure in the same area is not covered in this book. This subject will be briefly revisited in Chapter 6 where we discuss the perspectives associated with the work presented in this book.

1.3.3. *Communications security and auditing*

Despite the performance problems mentioned (section 1.3.2), the ad hoc V2V mode of communication, more generally the direct communication between terminals (*device-to-device* (D2D)), has been integrated into the architecture of communications network infrastructures, which includes 5G. This choice seems to confirm the usefulness of this mode at least for some specific use cases that are important enough to be taken into account. Nonetheless, contrary to the conventional understanding of ad hoc mobile networks where participants can freely join and leave the network, as long as they are equipped with a compatible wireless interface, the 5G V2V and D2D modes require each terminal to be previously authenticated with an operator before it is allowed to carry out exchanges in direct mode.

This is again an important issue that partly explains the difficulty of ad hoc communications to emerge in certain critical applications, including road vehicle networks. As a matter of fact, the emergence of communication systems dedicated to vehicles, driven or autonomous, allows for a variety of applications ranging from remote control to on-board leisure applications, including driver assistance systems designed to reduce road hazards. Some of these applications may have direct actions on the vehicle (emergency braking, trajectory control, etc.). In particular, emerging so-called cooperative applications enable vehicles to exchange information about their perception of their environment, their location and notifications of potential hazards encountered on the road (Rivoirard et al. 2018). Any V2V or V2I communication, or simply vehicle-to-everything (V2X), becomes a vector by which an actor, software program or human, can deceive the driver, human or software, of a vehicle with detrimental consequences for third parties. A vehicle application made available by an identified third party (car manufacturer through the on-board computer, a service platform through the smartphone or a connected box provider) usually relies on V2I communications. The information provided to the vehicles is guaranteed by the identified third party. In the event of damage, the technical means for investigation are provided for by the audit procedures of the car manufacturer, the telecommunications operator or by the third party identified as the application publisher. On the other hand, for so-called cooperative applications, the service is developed and provided based on direct exchanges between vehicles by way of V2V communications. Every vehicle potentially becomes a service provider or consumer with respect to this application. These exchanges, which are not controlled by a regulator, decrease the ability to identify third parties that could disrupt traffic. The lack of an identified third party centralizing and verifying the consistency of the data disseminated by the application increases the risk of spreading false information.

Several studies address this problem and propose solutions for the distributed management of identification for authentication, anonymity (via pseudonyms in particular) and reputation in order to assess the credibility of notifications. Similarly, for previously identified vehicles, in particular with a 5G infrastructure, it is obviously possible to find the identity of any vehicle that has sent a notification. However, when the exchange is carried out through an ad hoc communication, and has therefore not passed through the operator's equipment supposed to keep track of it, how can the content that has actually been exchanged be found in the event of an audit? Many studies have recently been proposed in order to create forgery-proof distributed ledgers, in particular through the use of specific *blockchains*, in order to organize the preservation of the trace of V2V or more broadly D2D exchanges.

These questions are irrelevant with respect to the objectives of this book, but they highlight issues that partly explain the fact that ad hoc communications did not emerge earlier in real-life concrete applications for vehicular networks.

1.3.4. *Technological evolution and depreciation*

Contrary to the remarks made about the rail network, the evolution of vehicles in road traffic places them for the most part in areas covered by the network infrastructure for the general public. It should also be noted that most of the applications for drivers were first provided through their smartphone or through the on-board computer of their vehicle, which is now mostly equipped with the same operating systems as smartphones. Intuitively, it can be said without taking too many risks that the technological evolution of network equipment embedded in vehicles will naturally follow that of consumer equipment, which in turn accompanies that of the network infrastructure of the operators. As a result, depreciation management will naturally follow that of the operator, which is clearly guaranteed as one of its priorities. It is therefore unlikely that this will lead to developing road vehicular networks with the technological gap observed today between the infrastructures of the network of operators (4G/5G) and the dedicated infrastructure deployed in the rail context (GSM-R). The involvement of the rail sector in a shared infrastructure, as long as guarantees about safety and security are given, in particular through guaranteed capacity and a complete separation between its dedicated network and the general public network within the shared infrastructure, would allow it to benefit from the same advantages in terms of evolution and depreciation.

1.3.5. *European harmonization*

Following the harmonization of European national rail traffic management systems through ERTMS, approximation has been initiated and certain phases are already operational in the harmonization of vehicle identification and information exchange systems. Insofar as the movement of vehicles is free between the different European territories and that the transfers of mobile device technology between European operators are now virtually transparent and at no additional cost, it is conceivable that applications and services provided from a territory can be ensured in the same way after crossing a European border.

Several European projects have been focusing on tasks that include this harmonization of services and applications, first between the various motorway operators and French car manufacturers, and then in larger consortia at European level. This subject also falls outside the scope of this book, but mentioning it makes it possible to bring forward the need for such considerations in the design and evaluation of a future communication system for all land transport.

1.4. Conclusion

Throughout this short introductory chapter, we have identified some of the issues surrounding ad hoc communications for the road. Despite the fact that some of them are not addressed in detail because they are unrelated to the scope of the objectives of this book, their mere mention facilitates a partial understanding, on the one hand, of the interest and, on the other hand, of the difficulties that arise by the adoption of ad hoc communications in effective vehicular networks despite a research background of a quarter of a century.

The experiment of communication networks for the railways, as the first full-scale experiment on a scale comparable to the road network in a given area, has shown some facilities; this is the case in decision-making in particular, which at the same time underlines the extent to which it could otherwise be more complicated in the context of road vehicle networks. On the other hand, it has also resulted in identifying the advantages of these networks, which should more easily integrate a shared infrastructure with other consumer services, better support changes and incur fewer costs.

Of all of the aspects developed in this chapter on ad hoc communications, the only aspect that will be developed in detail is the self-organization of vehicles and communications in order to be able to support cooperative applications when no communication infrastructure covers the area where the vehicles are traveling.

1.5. References

Ai, B., Molisch, A.F., Rupp, M., Zhong, Z.-D. (2020). 5G key technologies for smart railways. *Proceedings of the IEEE*, 108(6), 856–893. doi: 10.1109/JPROC.2020.2988595.

Pinedo, C., Aguado, M., Lopez, I., Astorga, J. (2015). Modelling and simulation of ERTMS for current and future mobile technologies. *International Journal of Vehicular Technology*, 912417. doi: 10.1155/2015/912417.

Rivoirard, L., Wahl, M., Sondi, P., Gruyer, D., Berbineau, M. (2018). A cooperative vehicle ego-localization application using V2V communications with CBL clustering. *2018 IEEE Intelligent Vehicles Symposium*, 4, 722–727. doi: 10.1109/IVS.2018.8500688.

Sniady, A. and Soler, J. (2014). LTE for railways: Impact on performance of ETCS railway signaling. *IEEE Vehicular Technology Magazine*, 9(2), 69–77. doi: 10.1109/MVT.2014.2310572.

Sondi, P., Ramat, E., Berbineau, M. (2017). A virtual laboratory as an assessment tool for wireless technologies in railway systems. In *Networking Simulation for Intelligent Transportation Systems*, Hilt, B., Berbineau, M., Vinel, A., Pirovano, A. (eds). ISTE Ltd, London, and John Wiley & Sons, New York. doi: 10.1002/ 9781119407447.ch5.

2

Functional Approach to Protocol Design

2.1. Introduction

Historically, driver assistance systems were developed earlier in guided transports (trains, metros and trams), which operate on segments of track equipped and almost exclusively intended for their use, unlike the automobile, which relies essentially on the driver's ability to control the vehicle. The increasing accessibility of signaling, perception, localization and communication systems in the automobile industry has allowed for the development of driver assistance systems, hereafter referred to as ADAS (advanced driver-assistance systems). These on-board systems and applications make it possible to respond to road safety challenges by reducing risks and mobility issues through the optimization of traffic and the minimization of congestion, as well as environmental problems through the optimization of fuel consumption and the reduction of pollution.

Generally speaking, all of these ADAS create a dependency in drivers who, in the long term, will have difficulty in facing road traffic without this assistance. Most people who have learned to drive in a recently-made car would find it very difficult to travel a distance of 200 km from the usual place (i.e. where they live) using an old car unequipped with braking assist, power steering or GPS. The development of new ADAS that rely on accessing a communication network, particularly, for example, remote vehicle control or centralized autonomous piloting applications, therefore necessarily implies that on roads located in areas in which there is no coverage for communication infrastructures, drivers will face problems. In addition, along with the communicating vehicle, so-called cooperative applications are being developed. For example, in case of cooperative localization or ego-localization, each communicating host vehicle utilizes the information received from neighboring vehicles to fine-tune and improve its own location. Cooperative extended perception

Design and Evaluation of Ad Hoc Routing Protocol,

operates on the same principle relatively to the sharing of electronic environment maps, hereafter referred to as LDM (*local dynamic map*).

Whether they are used in manned or autonomous vehicles, all of these multi-sensor systems, such as the aforementioned ADAS, cannot exclusively rely on communications involving an infrastructure, which is known as *vehicle-to-infrastructure* (V2I) communications. Several models recommend the development of ad hoc communications solutions as an alternative or in addition to the infrastructure. Besides the fact that there are places where the infrastructure may not be available or temporarily malfunctioning, the latter could also be ineffective in the presence of occasional degraded conditions. While the advantages of a centralized approach are its accuracy and capacity, it requires radio access points at regular intervals. However, this cannot be guaranteed for the entire road network, mainly for reasons of return on investment, especially in road sections that are generally less frequented. In addition, information transmitted through the infrastructure can present a risk of increased obsolescence compared to direct vehicle-to-vehicle (V2V) communication when the vehicles involved are in close proximity. Finally, the implementation of certain cooperative application functions through V2V communications would contribute to the overall decrease in the load of the infrastructure, which would enable us to manage a higher volume of traffic and applications to be more flexible.

However, despite these arguments that support devising an alternative network organization based solely on V2V communications, their operational implementation is not straightforward. This is because distributed media access and routing algorithms in ad hoc networks are more complex and their load on traffic further increase that of the applications supported by the network. It is therefore necessary to develop routing protocols that can be adapted to this very specific road context, but that are sufficiently efficient so that a greater proportion of resources can be allocated to the routing of application traffic.

After a brief overview of the basic concepts of routing and protocol engineering, this chapter presents a functional approach to designing an ad hoc routing protocol for the road environment. It draws on a short literature review on the major trends in routing in ad hoc networks to illustrate how the functional requirements of an ad hoc vehicular network can be used to express routing protocol design objectives and orientate its implementation. A case study is then proposed through the design of a recently proposed organizational model for ad hoc routing protocols in a road environment.

2.2. Ad hoc routing protocol basics

2.2.1. *Routing definition*

In computer terminal networks, to which on-board vehicle computers can be compared, routing refers to the process of packet or datagram transmission from a source terminal to one or more destination terminals. In the open system interconnexion (OSI) reference model, routing operations are handled at the level of layer 3 (the "network" layer). At the same level, an addressing mechanism, in this case IP addressing on the Internet, makes it possible to clearly identify the destination nodes. For example, Figure 2.1 shows that it is possible to route a datagram from N_i to N_9 through several routes, including the one shown in red as well as the one in green that will actually be chosen for some reason by the applied routing algorithm.

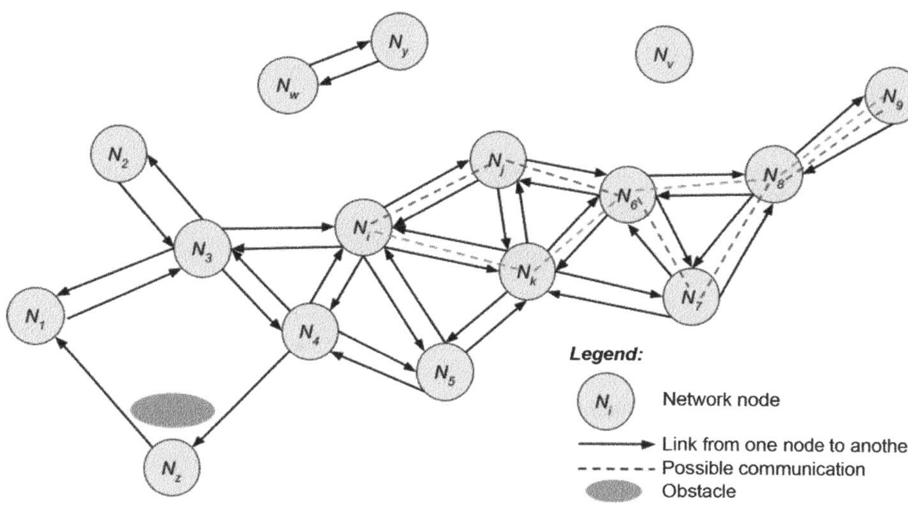

Figure 2.1. *Example network to illustrate the basic notions. For a color version of this figure, see www.iste.co.uk/wahl/design.zip*

The implementation of routing therefore requires an explicitation of the identification of the destination nodes, the mechanisms for discovering the possible routes, the criteria guiding the selection and the way in which this choice is implemented at the level of all of the nodes involved from the source to the destination. In this study, we deliberately focus on level 3 of the OSI layer and therefore ignore all of the physical phenomena observed at the level of layers 1 and 2 (respectively, the "physical" and "data link" layers). The only elements on which choices, their implementation and the resulting interpretations depend are the reception and processing of IP datagrams.

2.2.2. *Destination node identification*

When a communicating node transmits or retransmits datagrams, they are broadcast into the environment of the node as radio waves. The latter can therefore naturally be perceived by nodes that are not destination nodes, but which evolve in the route environment by which these datagrams are transmitted. The identification of the destination node(s) responsible for receiving and processing the datagrams, beyond simple perception, is based on the addressing system of the network layer. In the context of wireless communications, this addressing system defines several communication modes (Figure 2.2) that allow for finely managing the structure of potential destination nodes: *unicast*, *broadcast*, *multicast* and *anycast*.

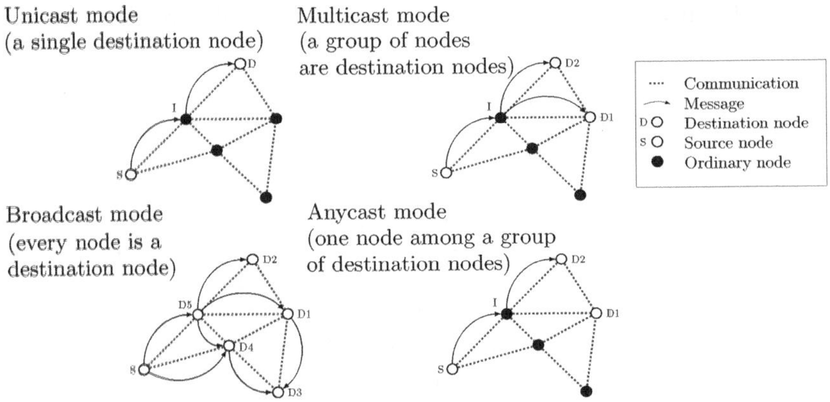

Figure 2.2. *Communication modes unicast, broadcast, multicast and anycast*

In the *unicast* communication mode, datagrams are transmitted to an address that identifies a single interface belonging to a single node. If this destination node is out of radio range of the sending node, one or more relay nodes must retransmit the datagrams until they reach the addressed node.

In the *broadcast* communication mode, datagrams are transmitted to all nodes in the network. Each receiving node in turn rebroadcasts them so that they reach all of the nodes in the network. The way in which this diffusion is achieved has a different impact on the network, hence the existence of a large body of work on adapted organizational models.

In the *multicast* communication mode, datagrams are transmitted to a group of nodes. When the destination nodes are out of radio range of the sending node, one or more relay nodes retransmit the packets until they are within range of the destination nodes. The IP addressing system provides ranges of addresses reserved to the identification of *multicast* groups. Nonetheless, when the group of destination nodes is defined according to the geographic location, this mode of communication is referred to as *geocast* (ETSI 2011a, 2011b). In this specific context, it may not be necessary to identify destination nodes by an address, since it is the geographical location that determines the group.

In the *anycast* communication mode, the emitted datagrams are destined for a single node chosen from a group of receiving nodes. When membership to this group is defined according to the geographic location of the nodes, this communication mode is referred to as *GeoAnyCast* (ETSI 2011a, 2011b). When the destination node is out of radio range of the sending node, one or more relay nodes retransmit the datagrams until they are within range of the destination node.

2.2.3. *Route discovery and selection*

Route discovery takes place in two main stages, namely, topology discovery and routing algorithm.

Apart from specific applications where the nodes forming the network can benefit from a prior exhaustive knowledge of the topology, the latter is most often discovered dynamically by each node as it evolves in the space covered by the network nodes. Initially, the discovery by a node is carried out through sent and received messages in its immediate environment. Some aspects of this discovery process is be illustrated in Figure 2.1.

By remaining in the network layer in the interpretation of this figure, node N_i receives the datagrams from node N_4. The latter also receives the datagrams of the previous one. The communication link between these two nodes is then said to be symmetrical. On the other hand, although node N_z receives datagrams from node N_4, its own transmissions, which do reach node N_1, are not received by the node N_4 due to an obstacle. It is said that node N_z appears hidden to node N_4 and the link from N_4 to N_z is asymmetric. Node N_v, which has no connection to any other node in the network, is said to be an isolated node. The nodes N_w and N_y, connected through a symmetric link to one another, form a connected component isolated from the rest of the network.

At this stage of route discovery, each node can only have knowledge of its direct neighbors from which it actually receives the datagrams. One way for a node to acknowledge more distant nodes is through retransmission. Looking at Figure 2.1, when node N_4 receives a datagram originating from node N_3, whose source is node N_2, knowledge of the number of retransmissions of this datagram suffices to deduce information about the topology. This is possible in IP networks through the TTL (*time-to-live*) field in the header of IP datagrams. In this case, since the message was retransmitted only once by node N_3, node N_4 can deduce that it is two hops away from node N_2. By generalizing, it is thus possible to deduce each time a datagram is received, how far the source is located in terms of number of hops and through which direct neighbor its datagrams reach the destination node. However, due to the existence of asymmetric links in a wireless communication network, this discovery does not guarantee that it will be possible to respond to the source through the same neighbor. Figure 2.1 actually shows that N_1 can receive datagrams from N_4 via N_z, but it is not possible for datagrams originating from N_1 to reach N_4 through N_z, because the latter does not receive messages from N_1 due to the obstacle.

The updating of topology information based on receiving and retransmission of messages depends on the frequency of occurrence of these events. For example, a node that does not receive or retransmit messages from certain parts of the network, such as its immediate neighborhood, will ignore the topology of the network for as long as it lasts. This is particularly the case for the isolated node N_v (Figure 2.1), which ignores the entire network topology, given that it does not receive any messages. In order to control the refresh rate of topology information, it is possible to set up dedicated messages for this purpose, usually referred to as control messages. Therefore, if each node sends a control message to the entire network every 2 s, its direct neighbors, which actually receive that message, will know every 2 s that it is part of their neighborhood. In addition, through retransmissions, more distant nodes will be able to locate it and know the number of hops in their own view of the network topology. The use of control messages therefore makes it possible to make route discovery independent of the frequency of datagram exchanges linked to application flows between nodes, or even to impose a periodicity on this discovery according to the field of application. For example, in a network of fixed sensors, this periodicity may be extensive, because the change in topology is more likely to occur when sensors fail, in particular due to battery depletion. On the other hand, in a network of vehicles traveling at relatively high speeds, it is necessary to set a very short periodicity without which only a permanently obsolete view of the network topology will be available. However, it should be kept in mind that the routes discovered by a node through these mechanisms are actually discovered from its point of view as a receiver. At this stage, there is therefore no guarantee, due to the existence of asymmetric links, that it will be able to use them as a sender when forwarding datagrams to the nodes discovered in the topology. If we reconsider the previous example (Figure 2.1) of node N_1 that would like to reach node N_4, it is easy to observe that the discovery of a good route and the exclusion of bad routes can be achieved simultaneously through two simple mechanisms.

The first mechanism consists of keeping only strictly symmetrical links in the topology view. As a result, the links N_4–N_z and N_z–N_1 are not preserved. One of the routes easily found in this case will be N_1–N_3–N_4. More generally, in a network where all nodes only keep the symmetric links in their topological view and only use them to transmit or retransmit datagrams, any route discovered by a receiver node can also be used by the receiver node to transmit to the initial source node. The establishment of the symmetrical nature of the link is therefore fundamental to this approach. Therefore, when node N_1 receives a control message sent from N_3, it discovers the asymmetric link from node N_3 to node N_1. However, at this point, node N_3 is not aware of any links, until it receives a control message from node N_1 to discover the asymmetric link from node N_1 to node N_3. From the point of view of the outside observer, there is now a symmetrical link existing between nodes N_1 and N_3, but from the point of view of each of these nodes, there is only an asymmetric link from the other node. There are two possible ways to reveal the symmetry of the link. The first method is a test that consists of sending, for example, for the node N_1 a specific request to node N_3. Upon receipt of a specific response to this request from node N_3, node N_1 deduces that the link to node N_3 exists, in addition to the link from node N_3 that it already knew, and can thus conclude that the link is symmetry. This first method therefore requires additional messages. Another, faster method involves including the list of neighbors that were already discovered when the control message was issued. Therefore, if after receiving the control message emitted by node N_1, node N_3 in turn sends its own, indicating that it has node N_1 in its neighborhood, the latter can immediately deduce that there is a link from node N_3 to it, because it receives its message, but also that there is a link from it to N_3 since the latter broadcasts that node N_1 is its neighbor (Figure 2.3).

Regardless of the symmetric nature of the links, the second mechanism consists of actively searching for routes to the targeted destinations. Rather than sending control messages to signal themselves to the network, and thus discover other nodes through their own broadcasts, each node uses control messages to ask the other nodes if they know a particular route for a specific destination. In this way, route discovery, which is now carried out from the point of view of the source node, is achieved the other way around compared to the discovery method described earlier. The routes obtained can therefore be used for datagram transmission, regardless of the existence of asymmetric links. On the other hand, this mechanism requires the implementation of specific control messages to transmit, on the one hand, the requests for routes and, on the other hand, the answers about the existence or the lack of routes. For instance, if node N_1 proceeds to an active search for a route to node N_4, its request reaches N_3, but not N_z because of the obstacle (Figure 2.1). Consequently, the only direct neighbor of N_1 that will have forwarded this request and will be able to respond to N_4 is N_3. The route N_1–N_3–N_4 is thus found again.

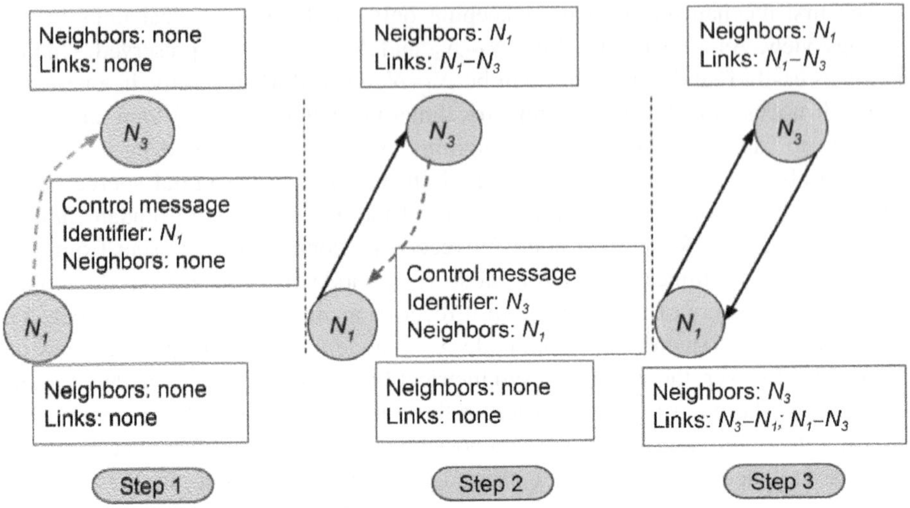

Figure 2.3. *Formation of a symmetric link as seen by node N_1. For a color version of this figure, see www.iste.co.uk/wahl/design.zip*

This same example highlights the general problem created by using an asymmetric link: the only node that knows about it is the receiver, and never the one that transmits. This is because node N_1 is aware of the asymmetric link N_z–N_1 as it receives datagrams originating from N_z. On the other hand, node N_z is completely unaware of the existence of any direct link to N_1 since no datagram from N_1 reaches it. Node N_1 is therefore the only one of the two that can take the initiative to use the asymmetric link to communicate with the other. As a result, as such:

– either the asymmetric state of the link is transient, until a possible first response from N_1 reaches N_z and a subsequent transmission of the latter enables N_1 to discover that a symmetric link exists between them. In this case, when N_1 decides to use the asymmetric link, it assumes that the status of the latter will change to a symmetric link shortly and that there is therefore a 50% chance that the transmission will succeed;

– or the asymmetric state of the link is rather permanent, especially if we consider the example where the obstacle is a long-term cause. From the point of view of N_1 that does not see the obstacle, there is always a 50% chance that the transmission will succeed. From the point of view of the outside observer, in this case the reader, it is clear that there is no chance of this happening.

The use of an asymmetrical link is therefore somewhat a gamble that the receiving node of this link takes as part of its approach to communicating with the source. Under certain conditions, such as the mobility of network nodes, which in fact makes any influence of fixed obstacles transient, or the general poor quality of multi-hop routes based on symmetric links, it is possible to consider the benefit/risk ratio related to the use of asymmetric links.

2.2.4. *Network topology discovery*

Regardless of the method chosen in section 2.2.3, through discovery or active search, the *broadcast* mode is used to disseminate the control messages to the entire network. The most basic method is conventional flooding or *blind flooding* (Lim and Kim 2001), in which each network node that receives a datagram for the first time systematically retransmits it in turn in broadcast. Therefore, the control messages emitted by a node N_i will be first retransmitted by all of the nodes in its neighborhood at a distance of a single hop (namely its neighbors within direct range of communication), denoted by $V_1(N_i)$, and then by the neighbors of the latter that are not themselves direct neighbors of N_i, and so on until, gradually, all of the nodes of the network have been reached. Nevertheless, in an ad hoc network, the network nodes share the resources of the same communication medium. In order to avoid collisions and interference between concurrent transmissions, when sending a packet by a source node, all transmissions from its neighboring nodes need to be stopped at the MAC layer. Flooding broadcasting then ensures reasonable performance for a communication network consisting of a very small number of nodes, having relatively little information to transmit and a low frequency of data transmission. As soon as the number of network nodes, or the amount of information transmitted by each node, or even the frequency of datagram transmissions and retransmissions sharply increases, this leads to a phenomenon called "broadcast storm". The resources of the communication medium rapidly reach saturation and no transmission becomes possible. Most datagrams retransmitted in such a context are useless as they will probably not reach the destination nodes, but in the meantime, they will still have contributed to further degradation of the network. *Blind flooding*, which overconsumes limited network bandwidth resources, is therefore not suitable for wireless ad hoc networks involving a large number of nodes and high-traffic applications such as vehicular networks. One solution, therefore, consists of limiting the discovery of the network topology carried out by each node to a horizon defined around the node that limits the total size of the network in its local view. If the discovery of the network around N_i (Figure 2.1) were to be limited to its neighborhood to a single hop, for example, the network topology would only comprise six nodes (N_i, N_3, N_4, N_5, N_j and N_k). Generally speaking, the strict neighborhood within *n* hops from any node N_i is defined as a set of the single-hop neighbors of the nodes that are comprised within strict neighborhood of the *n-1* hops from N_i, excluding N_i as well as any nodes that the node can reach through a path with fewer than *n* hops. This is summarized by the following equation:

$$V_n(N_i) = \left(\bigcup_{y \in V_{n-1}(N_i)} V_1(y) \right) - \left(\bigcup_{p=1}^{n-1} V_p(N_i) \right) - \{N_i\}, n > 1 \qquad [2.1]$$

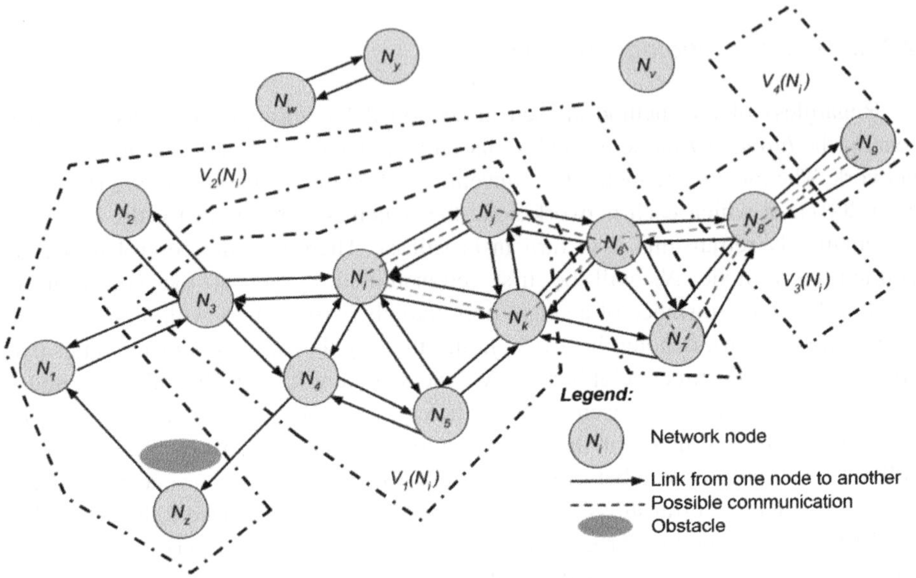

Figure 2.4. *Illustration of the one-, two-, three-, and four-hop strict neighborhoods of node N_i. For a color version of this figure, see www.iste.co.uk/wahl/design.zip*

From any arbitrary node N_i, the network topology can therefore be seen as a succession of its strict neighborhoods within one, two and so on, up to *n* hops. It is therefore possible to restrict to a given number of hops the neighborhoods to be included during network discovery according to a critical preset size, as the number of nodes, with regard to message sizes and their transmission frequency by the targeted applications. A solution such as the *Fisheye* extension works on this principle (Guangyu Pei et al. 2000). It is based on the idea that the further away a neighborhood is, the less accurate the information available about the nodes comprised therein, in particular because they take longer to reach the source node, which increases their risk of obsolescence. Consequently, the amount of information to be aggregated and transmitted, the algorithmic effort involved in finding optimal routes, and the frequency of control messages should be reduced as the neighborhood in question is farther away from the source node. Therefore, some protocols are based only on exhaustive knowledge of the single-hop neighborhood, while others regularly maintain their topology up to two hops or more, for example, in inverse proportion to the mobility of the nodes. The faster and more often the nodes move, the more frequently it is necessary to refresh the topology knowledge. However, due to overlapping topological views between different nodes in the network, the effectiveness of the ad hoc routing protocol at the global network level requires other network partitioning strategies.

2.3. A few strategies adopted by ad hoc routing protocols

A routing protocol is a specific piece of software that is executed by each node in the network and whose purpose is to precisely define the information exchanged and the rules applied at all levels during routing operations. For example, it defines the format of control messages, the conditions for using detected asymmetric and symmetric links, the routing algorithm to be applied and the policies for controlling the local view of the network topology. Since the objective of a routing protocol is to allow each node in the network to send datagrams to any destination node in its topological view of the network, the impact of all of the mechanisms related to its operation on communication resources must be limited in order to preserve them for the requirements of application traffic. Consequently, each routing protocol implements a set of strategies for the purpose of optimizing the dissemination of datagrams resulting from routing messages and application flows.

In order to better understand the strategies adopted by ad hoc routing protocols in the literature, their description can be based on a decomposition into two phases. The first phase involves the establishing of logical links and network topology. The second phase focuses on message dissemination when searching for routes.

2.3.1. *Protocol strategies for establishing links*

As discussed earlier in section 2.2.3, neighborhood discovery can be done by way of the periodic transmission of control messages, in this case HELLO messages (also called *beacons*). This neighborhood discovery step is the basis for forming the network topology as described in section 2.2.4. In the ad hoc routing protocols found in the literature, this topology can be horizontal or hierarchical.

2.3.1.1. *Horizontal topology*

Protocols adopting a horizontal topology do not confer a privileged function to any node: any node in the network plays the same role and functions as its neighboring nodes. This topology is the same as most ad hoc routing protocols, and in particular the *Destination-Sequenced Distance-Vector Routing* (DSDV) protocol (Perkins and Bhagwat 1994). Protocols that make use of a hierarchical topology assign different functions to certain nodes, thus building logical hierarchical links between nodes on top of the physical topology.

2.3.1.2. *Hierarchical multipoint relay mesh topology*

The technique of multipoint relays (MPRs) (Qayyum et al. 2000) creates a hierarchical topology by assigning a particular function to certain nodes of the network, which are the only ones authorized to retransmit the datagrams transmitted in broadcast originating from the nodes that elected them. Each node N_i in the network elects a minimal set of MPR relay nodes from the nodes within its single-hop neighborhood $V_1(N_i)$ that allow it to reach all nodes in its two-hop neighborhood $V_2(N_i)$. The set of all of the elected MPR network nodes forms a dominant group. This set thus formed in a distributed manner by local choices made at the level of each node is not minimal at the global level of the network. This is moreover not necessary (Gantsou et al. 2009).

Figure 2.5 actually illustrates the broadcast diffusion of a datagram by the multipoint relay technique compared to the flooding-based method (*blind flooding*). Only neighbors that have been elected MPR (in green) by the source node (in red) retransmit the datagram. Then, only the nodes elected as MPR (in blue) by the previous retransmitters (in green) retransmit it in turn, and so on until all nodes in the network have received the initial datagram. In other words, regardless of the total number of MPRs in the overall network, what's important is that the number of MPRs actually involved in the retransmission of any datagram transmitted via broadcast gets closer to the minimum necessary to reach all nodes in the network. A count, proposed in Sondi (2010), shows that it is sufficient to minimize the number of MPRs locally selected by each network node to obtain a number of retransmissions very close to the minimum possible. In the case of flooding-based routing, all nodes in the network rebroadcast the datagram. In this example, the total number of nodes relaying it in the three steps is nine for the multipoint relay method (Figure 2.5a), and 21 for the *blind flooding* method (Figure 2.5b). Broadcasting by the MPR technique can therefore reduce by more than half the number of relays involved in the broadcast of a datagram intended for the entire network, even for fairly low densities as in this example. The very good score obtained in this example is not completely coincidental, because the particular layout of the neighbors in successive concentric circles with respect to the source is well suited to the optimization obtained by the MPRs. As a matter of fact, Sondi (2010) shows that the spatial distribution of nodes, generated with the algorithm proposed by Rice (2006) according to a normal distribution that corresponds to this example, leads to more favorable results compared to a uniform distribution or a stationary distribution (Navidi and Camp 2004) resulting from a mobility such as *random waypoint*. Nevertheless, independently of the spatial distribution, the interest of MPRs is factual and increases with the node density in the network.

Due to its efficiency, the multipoint relay technique is used in particular in the *Optimized Link State Routing* (OLSR) protocol (Clausen et al. 2003).

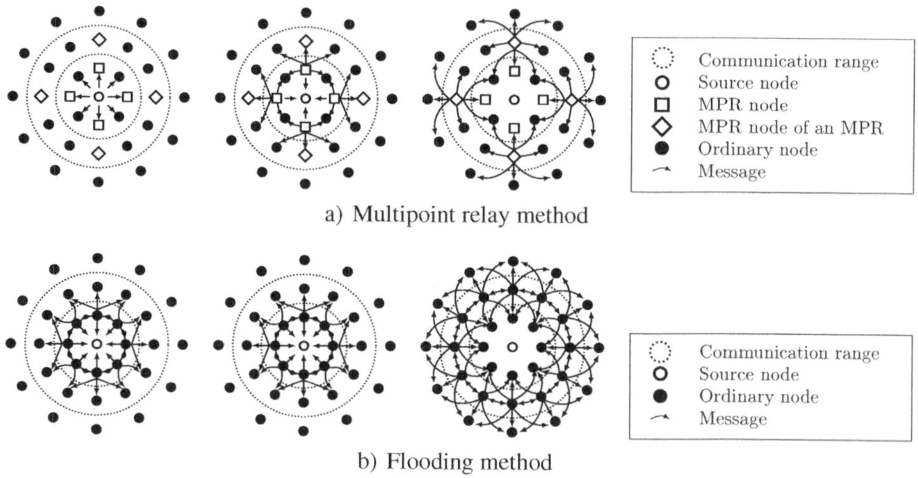

Figure 2.5. *Multipoint relay and flooding-based routing strategies (Rivoirard 2018)*

2.3.1.3. *Hierarchical topology by grouping or clustering*

Methods for distributing nodes into groups (*clustering*) perform a virtual slicing of the network based on a well-defined set of rules. When used in ad hoc networks, these methods aim to contain the broadcast of datagrams to one or more particular cluster, so as to reduce the number of unnecessary retransmissions. Clusters are formed around nodes that are usually referred to as cluster-heads (or group leaders). Other special roles can also be defined, each of which refers to different properties and functions. Vèque and Johnen (2012) list four node roles (Figure 2.6):

– ordinary nodes, which are each a member of a single cluster;

– the cluster-heads, which are the cluster-heads to which the ordinary nodes are attached;

– gateway nodes, that are members of more than one cluster; they allow for the transfer of information between the clusters of which they are members;

– undefined nodes that, at any given time, do not have any of the previously defined roles (ordinary, cluster-heads, or gateway).

The *clustering* methods for ad hoc networks have been the subject of much work, which has been listed in some previous studies (Bali et al. 2014; AlMheiri and AlQamzi 2015; Cooper et al. 2017; Senouci et al. 2020). They are categorized according to a variety of criteria, including:

– the criteria and metrics used for the formation and maintenance of the clusters, as well as for the election of their head;

– the number of hops that separate a regular node from its cluster-head;

– the dominant nature of the node clusters. A cluster can be a *dominating set*, a *connected dominating set* or a *weakly connected dominating set*. A cluster of nodes is dominating in a network if any node in the network is part of that cluster or is within a hop of a node that is part of it. A cluster is said to be connected dominating if the cluster is dominating and any in the cluster can reach any cluster node through a path whose intermediate nodes are in the cluster. A weakly connected dominating set is a connected dominating set with the lowest number of nodes among all the connected dominating sets in the network;

– the disjoint or overlapping nature of clusters: clusters are disjoint when there is no node that is simultaneously a member of two clusters (such clusters do not have gateway nodes); clusters are likely to overlap when nodes can be members of more than one cluster (there are gateway nodes);

– the active, passive, or hybrid nature of the *clustering* method: an active method transmits specific routing messages to form and organize clusters independently of the transmission needs of the applications; a passive method creates clusters only when a node has information to transmit; a hybrid method includes in frequently sent routing messages information necessary for cluster formation and maintenance;

– the number of different hierarchical levels (clusters, subclusters, and so on) created by clustering.

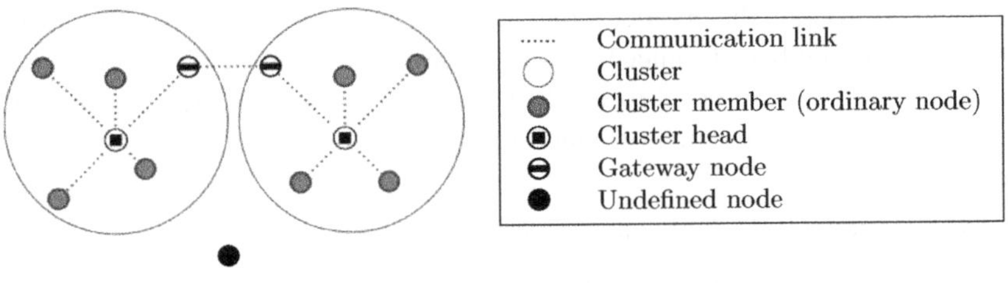

Figure 2.6. *Node functions in a cluster organization*

The advantage of clustering methods is that they facilitate scaling up (Vèque and Johnen 2012). Since some communications can be confined within a cluster, the cluster-head can be used to coordinate internal communications and be in charge of forwarding outbound communications from the cluster on their own, thus reducing retransmissions in the network. This characteristic makes this approach promising for mobile ad hoc networks (MANETs) such as vehicular ad hoc networks (VANETs), where the number of nodes can significantly increase in response to road traffic. The disadvantages of *clustering* relate essentially to the routing traffic required to create and maintain clusters. More particularly, in order to maintain an up-to-date view of the network topology, routing messages should be sent more frequently when network nodes are highly mobile.

2.3.1.4. *Hierarchical backbone forming topology*

The logical network structuring around a backbone is based on the use of a clustering method. A clustering method identifies the node cluster-heads, and the *backbone* approach introduces a logical connection between these cluster-heads (Figure 2.7). In this type of organization, all nodes are within k-hops from a backbone node. This approach was particularly studied in Dagdeviren and Erciyes (2006), which proposed the creation of a backbone according to a circular ring topology.

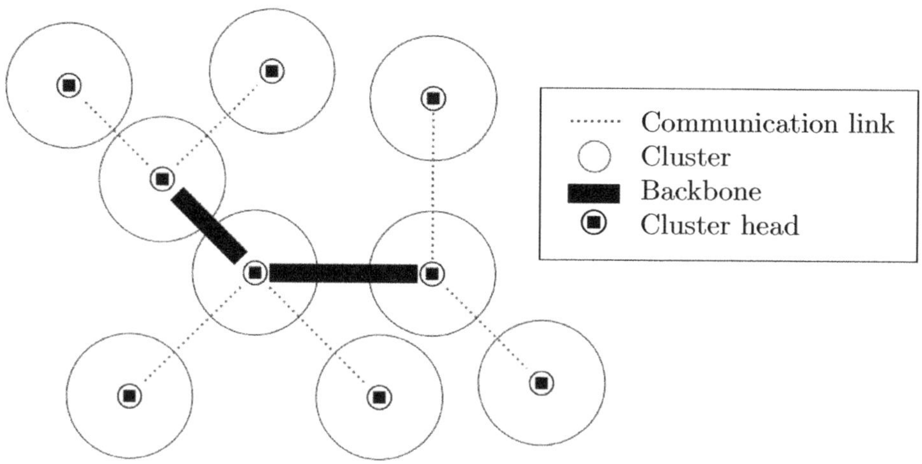

Figure 2.7. *Hierarchical backbone-type topology within an ad hoc network*

2.3.2. *Protocol strategies in request dissemination*

As described in section 2.2.3, the search for paths is performed by disseminating control messages, which may contain routing requests, among other things. The purpose of a routing request is to identify all of the intermediate nodes (relay nodes) that will relay the datagrams related to the application traffic of a sending node to one or more destination nodes, if the destination node or destination nodes are out of direct radio range of the source. There are three main methods of disseminating request. Proactive, reactive and hybrid methods are characterized by the choice of the moment when the search for routes is initiated (Olariu and Weigle 2009).

2.3.2.1. *Proactive approach*

In a proactive approach, each node maintains a routing table, in addition to its neighbor tables up to k hops. The value of k varies between protocols, most often $k = 1$, but in the OLSR protocol, for example, k is 2. The node periodically broadcasts routing messages across the network, independent of any request to transmit application traffic. These control messages contain a list of routes known to the node. These exchanges allow the management of routing tables. A routing table contains at least the address of a potentially destination node and the address of the first relay (one-hop) capable of rebroadcasting the information to this destination node (Figure 2.8). Some protocols associate these two pieces of information with a cost characterizing the related the path. An example of a cost metric is the number of relay nodes needed to reach the destination node (number of hops) via the path in question. For example, during a *unicast*-mode transmission request, a node forwards its datagrams to the address of the first relay node to the destination node. Upon receipt of each datagram, the relay node proceeds in the same way, retransmitting it to the relay node identified in its own routing table to the destination, and so on to the destination. DSDV (Perkins and Bhagwat 1994), OLSR (Clausen et al. 2003) and TBPRF (*Topology Dissemination Based on Reverse-Path Forwarding*) (Ogier et al. 2004) are examples of proactive protocols for MANETs.

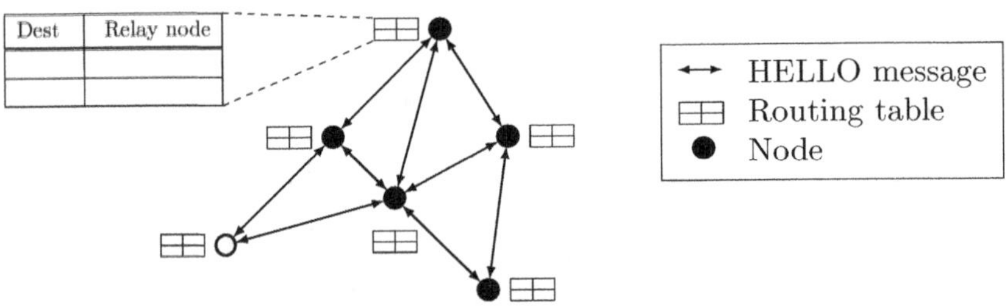

Figure 2.8. *Overall operation proactive methods. Figure inspired by Olariu and Weigle (2009)*

By periodically updating the routing table, proactive dissemination methods make available, at any time, a path to any node that can be reached in the topological view of the network from the host. The support of all application traffic is therefore almost immediate and explains the responsiveness and often lower delays of proactive protocols compared to other protocols. However, the management of routing tables requires the implementation of persistent periodic routing traffic that consumes network resources at the expense of application traffic. Nonetheless, this routing traffic can be completely anticipated and controlled through various protocol operating parameters. The idea is therefore to reduce it to the minimum necessary for an expected level of performance.

2.3.2.2. *Reactive approach*

In a reactive approach, path lookups are performed at each application packet delivery instruction using *Route Requests* (RREQ) and *Route Replies* (RREP) messages (Figure 2.9). An RREQ message is sent in *broadcast* mode and retransmitted from node to node until it reaches the destination node.

For the DSR (*Dynamic Source Routing*) protocol (Johnson et al. 2007), before retransmitting an RREQ, each intermediate node adds its address to it. The destination node then returns an RREP routing message to the source node containing the list of addresses of the intermediate nodes that retransmitted the RREQ message. This destination node can receive multiple copies of the RREQ message of which have followed different routes. The destination node can then select one of the routes to build the *Route Replies* (RREP) response message. Upon receipt of the RREP message, the source node reconstructs the route to be used for the application traffic datagrams to the destination node. This particular routing mode, where the route to be followed by the datagrams is set by the source, is called "source routing". It has the advantage of having a fixed path where all the intermediate relays are known, which can be interesting for many reasons, including safety. However, in the context of node mobility, a path built in this way is a rigid structure that can quickly become obsolete, as opposed to an approach where the source would trust the relays to adjust the path to follow as the datagram gets closer to the destination.

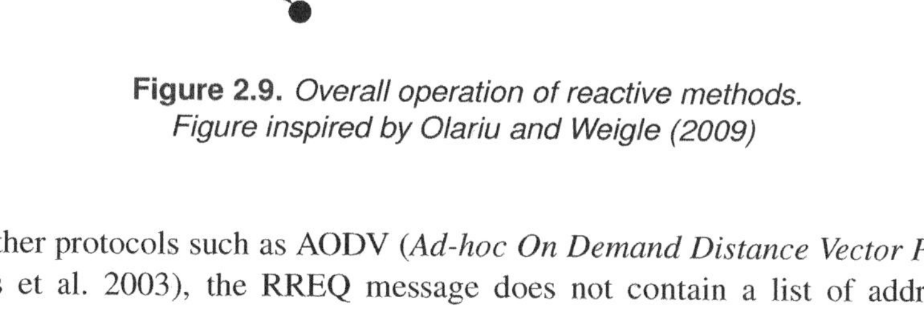

Figure 2.9. *Overall operation of reactive methods. Figure inspired by Olariu and Weigle (2009)*

In other protocols such as AODV (*Ad-hoc On Demand Distance Vector Protocol*) (Perkins et al. 2003), the RREQ message does not contain a list of addresses of the intermediate nodes. These intermediate addresses are locally stored in routing tables. Upon receipt of an RREP or RREQ packet, the nodes store a set of three addresses: the address of the source node that made the request, the address of the neighboring node that relayed the request and the address of the destination node. Based on these distance vectors, each node calculates its routing table so that the best path to the destination is available. This is referred to as "hop-by-hop routing".

In reactive protocols, a cost metric associated with routes can be considered when calculating routing tables.

Unlike proactive approaches, reactive dissemination methods do not require periodically sending control messages across the network for their operation. However, a path search delay is often observed at each application traffic emission processing, which delays the time when transmission can actually begin.

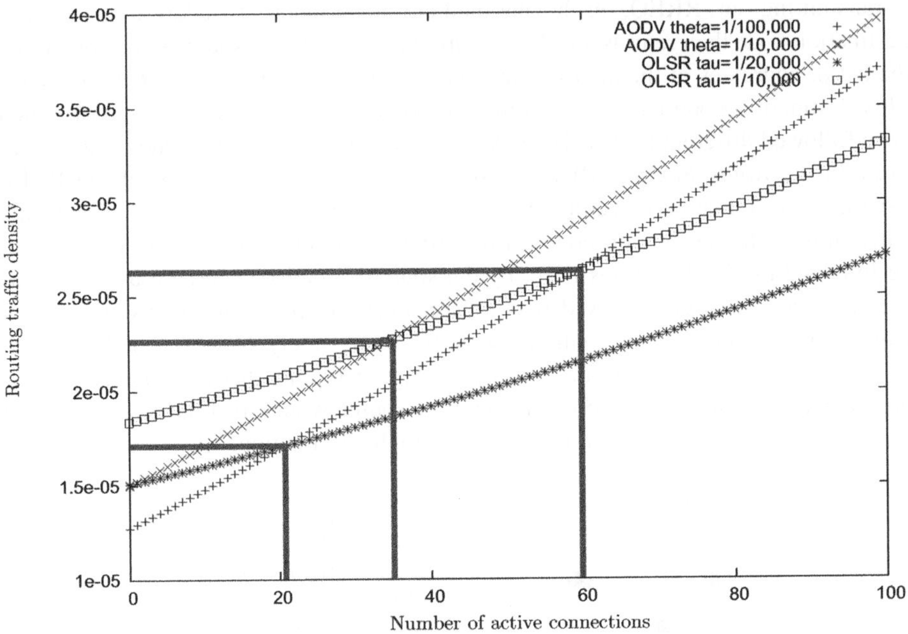

Figure 2.10. *Comparison of OLSR and AODV on routing traffic (source: Lebedev 2006). For a color version of this figure, see www.iste.co.uk/wahl/design.zip*

Figure 2.10 extracted from Lebedev (2006) and annotated by Rivoirard (2018) analytically compares the routing traffic generated by the OLSR proactive protocol and the AODV reactive protocol based on the number of nodes in the network for intrinsically identical parameters in the operation of these protocols. Only two parameters τ and θ are specific. The parameter τ quantifies the frequency with which OLSR topology control (TC) messages are sent, and θ quantifies the disruptions of RREP and RREQ requests of the AODV protocol. We observe stall points at which AODV routing traffic becomes higher than OLSR traffic from a certain number of nodes in the network, regardless of the parametrization of the two protocols. OLSR generates less routing traffic for a large number of nodes in the network for inherently equivalent parameters. This tends to confirm the general idea that the implementation of an organization prior to communications as is done for the OLSR proactive protocol is more suitable for high-density networks than a purely reactive behavior adopted by the AODV protocol.

2.3.2.3. *Hybrid approach*

The hybrid approach combines reactive and proactive approaches. A proactive approach is adopted in a region close to the node up to the neighborhood within a certain number of k hops from the host, and a reactive approach is taken for routing to nodes farther away (Olariu and Weigle 2009). According to this principle, Figure 2.11 shows a nearby region defined by a one-hop neighborhood around the source node S. On the other hand, the destination node D, two hops away, is outside of this nearby region. To find out the path to neighboring nodes in their near region, each node periodically transmits control messages. To reach nodes outside this region, a path search is triggered when application traffic datagrams arrive and is performed through exchanges of RREQ-RREP message. The ZRP (*Zone Routing Protocol*) (Beijar 2002) is an example of a hybrid protocol for ad hoc networks.

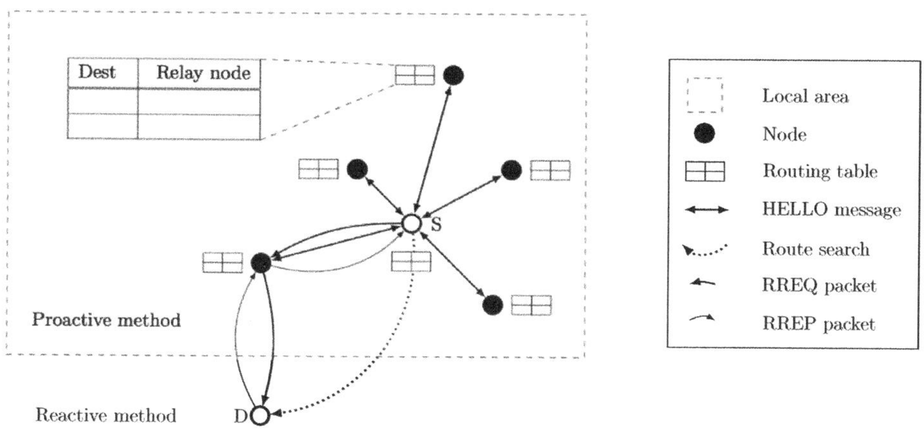

Figure 2.11. *Hybrid approach operation. Figure inspired by Olariu and Weigle (2009)*

2.4. Ad hoc routing protocols for road traffic

2.4.1. *Transposition of protocols from mobile ad hoc networks*

Historically, the proposed ad hoc routing protocols for MANETs have been employed for vehicular ad hoc networks (VANETs). While research on ad hoc vehicle networks was already underway, two new approaches were developed to complement the strategies described in section 2.3.

In the first approach, known as opportunistic (Chakchouk 2015) and from which the proposition (Li 2015) for VANETs is derived, the network nodes make the routing decision according to the state of the network and on the basis of predefined characteristics. This strategy operates on a horizontal topology and does not aim at building a hierarchical structure. Nevertheless, in design and during the evaluations carried out, it has been shown that it allows for a certain rate of connectedness between vehicles to be achieved according to different vehicle densities.

The second approach includes all of the so-called geographical protocols (Boussoufa-Lahlah et al. 2018), including the *Geographic Routing Protocol* (GRP) (Zhiyuan 2009). The latter utilizes a greedy routing method that locally elects, hop after hop, each relay node minimizing the distance separating it from the destination node, but in association with a hierarchical division of the operational space of the network into rectangular areas, subareas and sub-subareas (Figure 2.12). The first division of the network corresponds to the hierarchical level 1. Each resulting area is then divided into four subareas of hierarchical level 2 and so on. In order to be able to rely on the structure of the areas, each node must be able to know its position and rely on a global localization service that provides the position of all potential destination nodes. However, the implementation of this global location service is not clearly explained in the GRP specification.

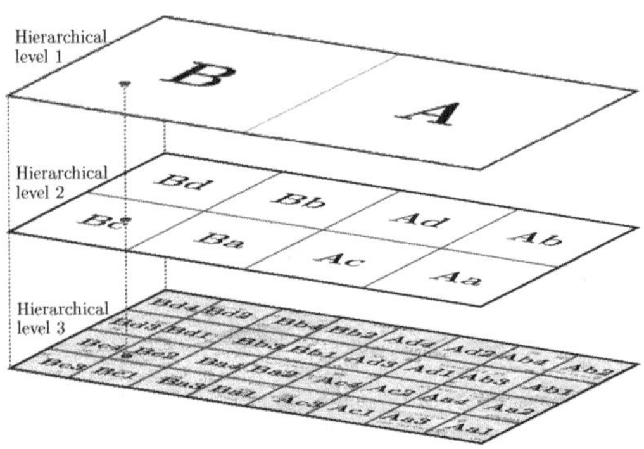

Figure 2.12. *Topological network division by the GRP protocol (source: Hussey et al. 2013). For a color version of this figure, see www.iste.co.uk/wahl/design.zip*

2.4.2. *Strategies and considerations for the road environment*

In contrast to the organizational models proposed for MANETs in general, which assume free mobility in an open space, VANETs are spatially restricted by the configuration of traffic lanes. As such, if the basis of the study is reduced to any segment between two intersections, the VANET evolves within a longitudinal space comprising one to several traffic lanes and one to two directions.

Vehicle-specific ad hoc routing strategies are therefore developed according to specific characteristics of the road environment (see Figure 2.13), namely:

– the internal vehicle characteristics, such as, for instance, position, speed or acceleration;

– the road environment characteristics external to the vehicle such as the type of road, the number of lanes, intersections and the road network;

– the hybrid characteristics resulting from vehicle(s)-environment interaction such as density and journey.

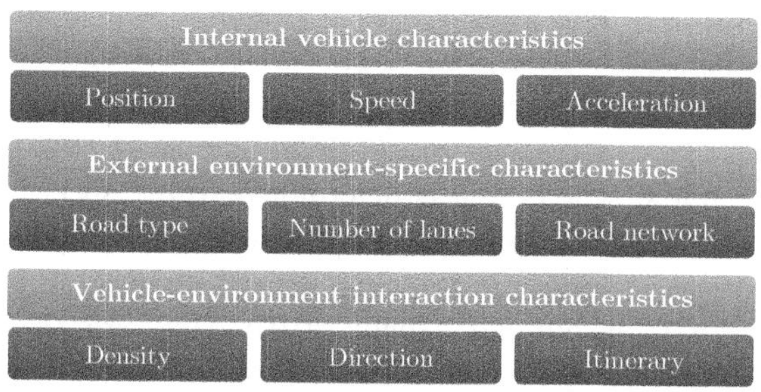

Figure 2.13. *Classification of information specific to the vehicular context (Rivoirard 2018). For a color version of this figure, see www.iste.co.uk/wahl/design.zip*

2.4.2.1. *Vehicle dynamics-related characteristics*

Vehicles operating in an environment covered by a communication infrastructure can obtain their position as well as that of their potential neighbors or destinations through a centralized localization service. This is the assumption already mentioned in the context of so-called geographical protocols such as GRP (Zhiyuan 2009). On the other hand, in a purely ad hoc context, vehicles must organize themselves to obtain this information without the need for any land infrastructure. Generally speaking, it is assumed that every vehicle can solve the problem of finding its own position via an on-board device using GPS or Galileo signals. On the other hand, the only way to obtain the position of neighbors and potential receivers consists of implementing the transmission of the position of each vehicle to the rest of the network through control or discovery messages. The methods listed by Ayaida et al. (2012) are categorized as flooding or "rendez-vous" methods. Flooding-based methods are used by proactive or reactive approaches in a horizontal topology. In rendez-vous based methods, certain nodes in the network are assigned a function of "localization server", de facto imposing a hierarchical topology.

Regardless of the method used, these exchanges of positions have led to the emergence of new cooperative applications, in which ego-localization can be found, enabling each vehicle to improve the accuracy of its own position based on that of its neighbors (Rohani et al. 2016). Several routing protocols employ the position information (Beylot and Labiod 2013; Blum et al. 2004; Boussoufa-Lahlah et al. 2018). A two-dimensional space is generally considered in most protocols in the literature, unlike the work of Lin et al. (2013) on the *Greedy Perimeter Stateless Routing* (GPSR) protocol evaluated in two- and three-dimensional spaces. Among other things, this position information makes it possible to do the following:

– Search for spatially constrained routes within a search area or relative to a source-destination search axis. This area can be rectangular as in the LAR (*Location-Aided Routing*) (Ko and Vaidya 2000) or triangular as in the DREAM (*Distance Routing Effect Algorithm for Mobility*) protocol (Basagni et al. 1998) (Figure 2.14). The initial search area is determined with respect to the source and destination positions, and then expanded progressively until there are enough relays to reach the destination. This search principle is used in an extension of AODV, the GeoAODV protocol (Hnatyshin and Asenov 2010). The CAR (*Connectivity-Aware Routing*) protocol (Naumov and Gross 2007) is based on the PGB (*Preferred Group Broadcasting*) method, which operates by searching for successive relays in a ring area defined according to the range (Figure 2.15). The GPSR protocol (Karp and Kung 2000) combines a perimeter-based method with a greedy algorithm (Figure 2.16) that gives preference to the selection of the relay node at the range limit and whose distance to the destination node is the shortest, with respect to the search axis defined by the positions of the source node and the destination node. Similarities with strategies adopted in traffic data sharing protocols can be seen (Delot et al. 2010).

– Select the cluster-head in the grouping methods (Maslekar et al. 2011; Tian et al. 2010; Di Felice et al. 2011; Xiaonan and Huanyan 2014). The *Position-Based Clustering* (PBC) protocol (Wang et al. 2008) uses the distance value to form groups that meet a previously set threshold. The *Cluster Chain* (CC) (Xiaonan and Huanyan 2014) and *Destination Based Routing* (DBR) protocols (Sethi and Chand 2017) elect the node located at the lowest average distance of all its neighbors as the cluster-head.

As with the position data, each node can broadcast its speed by way of control messages, more specifically HELLO messages, whose advantage is that they are periodically sent. It is possible to avoid this additional increased traffic created by the control messages by inferring the speed of neighbors from their successive positions. However, this would then represent an average speed over an observation time window, corresponding, for example, to the periodicity of HELLO messages, and which could be distorted both by too long a periodicity and by possible changes in the direction of those neighbors. Protocols in the literature thus either involve the relative node speed (Rawashdeh and Mahmud 2012; Vèque et al. 2013;

Togou et al. 2014; Daknou et al. 2015; Mehra et al. 2016), or the average node speed (Arkian et al. 2014; Azizian et al. 2016; Sethi and Chand 2017) and, in some cases, to the instantaneous speed (Wu et al. 2013a; Wu et al. 2016).

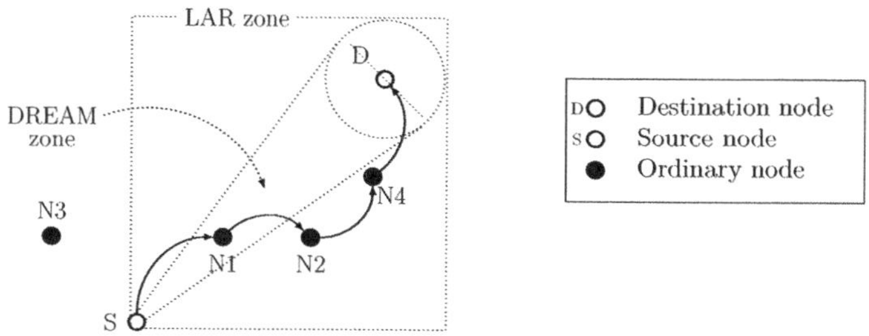

Figure 2.14. *Restriction of the path search using a geographic area*

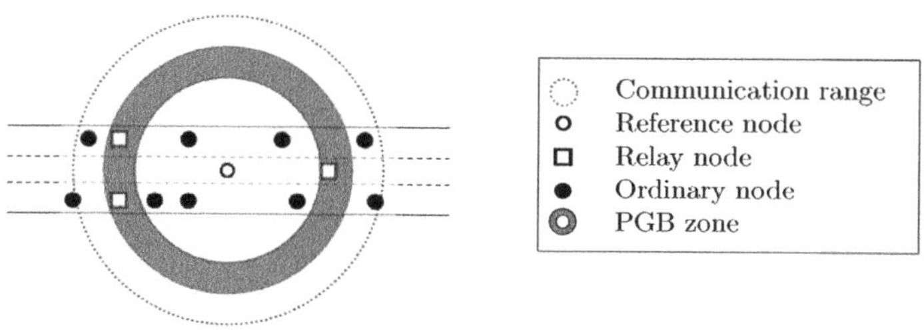

Figure 2.15. *Local geographic restriction of the PGB method. Diagram inspired by Mukunthan (2015)*

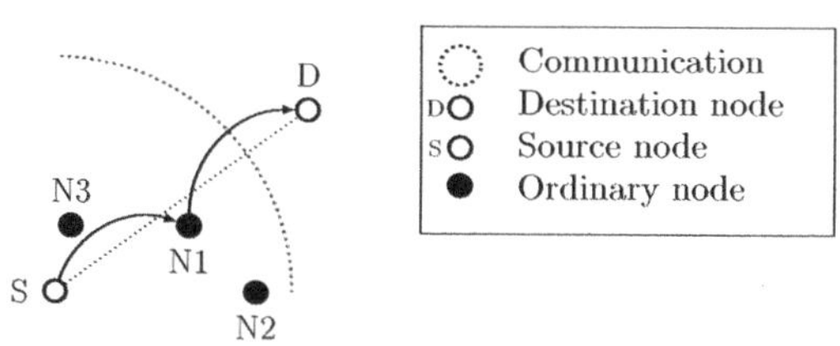

Figure 2.16. *Relay node chosen by the greedy routing algorithm*

Relative velocity is used when forming clusters of communicating vehicles with the aim of forming the most stable structure possible over time, especially in the context of the *Stable Cluster* (SC) (Rawashdeh and Mahmud 2012) protocol. In particular, the average speed is integrated into a score function that combines the position and quality of the links to elect the cluster-heads (Arkian et al. 2014). It is also found combined with mean relative distance as a mobility metric for forming adaptive size clusters as part of the *Distributed Cluster formation for VANET based on End-to-End Relative mobility* (DCEV) protocol (Azizian et al. 2016).

Some algorithms for computing *backbones* select the nodes with the lowest velocity among their neighbors to initiate the formation of the backbone. Instantaneous speed is also used by *BackBone Routing* (BBR) (Wu et al. 2013a), *BackBone Broadcast Routing* (BBBR) (Wu et al. 2013b), as well as their improved *Backbone Based routing with inter-flow Network Coding* (BBNC) protocols (Wu et al. 2016).

It was observed that acceleration had an impact on the quality of communications (Rivoirard et al. 2016). However, it is generally not used in a direct way. In particular, it is found combined with relative velocity in a metric called "spatial dependency" in the *Dynamic Clustering Algorithm* (DCA) protocol (Fan et al. 2011).

2.4.2.2. *Road environment characteristics*

The road network allows for a very wide variety of configurations, each requiring a particular management of the organization of communications between the vehicles in traffic in order for it to be efficient. We will return in more detail later to the motorway environment, which is the subject of the case study proposed in this book. Several protocols have been proposed for applications in urban environments, particularly those which use an anchor system (Chen and Gerla 1998; Seet et al. 2004; Jerbi et al. 2006).

Intersections therein are particular network points that are ideally positioned to give them priority for packet routing. The implementation requires each communication node to have access to accurate and up-to-date mapping of the road infrastructure in order to be able to more easily search for and detect these relays. The *Geographic Source Routing* (GSR) protocol (Chen and Gerla 1998) relies on the shortest path algorithm (Dijkstra) to form the list of intersections to be crossed in the direction of the destination node, and each relay node applies the greedy method to select the following relay (Figure 2.17). The *Anchor based Street and Traffic-Aware Routing* (A-STAR) (Seet et al. 2004) and *Greedy Traffic Aware Routing* (GyTAR) (Jerbi et al. 2006) protocols were subsequently defined based on GSR. A variation is introduced by the A-STAR protocol, which selects intersections located on the usual public transport routes, considered to have a stronger connection. The GyTAR protocol differs from the previous two by weighting the different intersections with the average vehicle density in adjacent streets to increase the likelihood of greater connectedness.

In urban areas, multi-lane roads are generally direction-selective at intersections (right lane to go right, middle lane to go straight or left, etc.). The *Clustering Based on Direction in Vehicular Environment* (C-DRIVE) (Maslekar et al. 2011) protocol is based on the detection of the vehicle lane to organize clusters upstream of an intersection. Several clusters can thus co-exist in a single street comprising several lanes of vehicle traffic. Similarly, the *Connected Dominating Set for Stable Virtual Backbone* (CDS-SVB) (Togou et al. 2014) protocol weights vehicles differently depending on the lane they are traveling in when electing cluster-heads, so as to penalize vehicles traveling in lanes that are more likely to lead them to an upcoming change of direction.

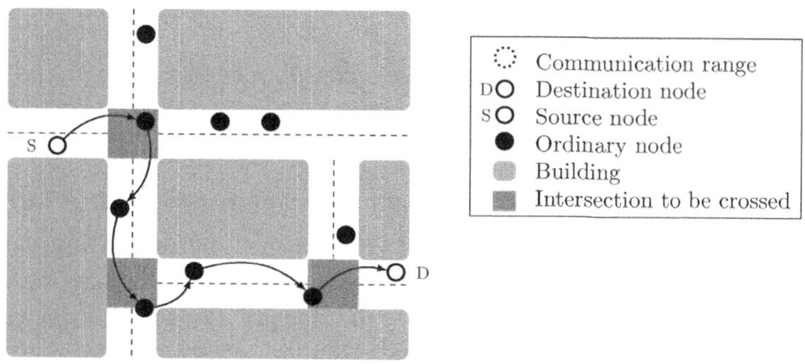

Figure 2.17. *Anchor system implemented by the GSR protocol (Rivoirard 2018). For a color version of this figure, see www.iste.co.uk/wahl/design.zip*

2.4.2.3. *Vehicle characteristics in road traffic*

Vehicle density on a road section is the measure of the average number of vehicles on that road divided by the cumulative length of the lanes in the said road section. In this specific case, this linear density is as relevant as surface density, since it is not normally possible for two vehicles to overlap on a single lane. Density can have a direct impact on wireless communications, as it determines, at least potentially, the number of source nodes in the same location and thus the level of competition for accessing the medium and the volume of data exchanged. In a wireless ad hoc network, apart from any specific application, nodes are deemed to enter and exit any road section without obeying any particular coordination rules. From a given node, an estimate of the local density can be obtained by dividing the average number of vehicles detected in its immediate vicinity by the product of the number of lanes and twice the range of the radio technology. This relationship can be generalized to the density measurement in an area including more distant neighborhoods as described in section 2.2.4. Figure 2.18 and equation [2.2] illustrate this measurement up to a neighborhood within n hops around a central node on a three-lane road section:

$$Density\ up\ to\ neighborhood\ V_n\ of\ N_i = \frac{1 + |\bigcup_{p=1}^{p=n} V_p(N_i)|}{3(2nR)} \qquad [2.2]$$

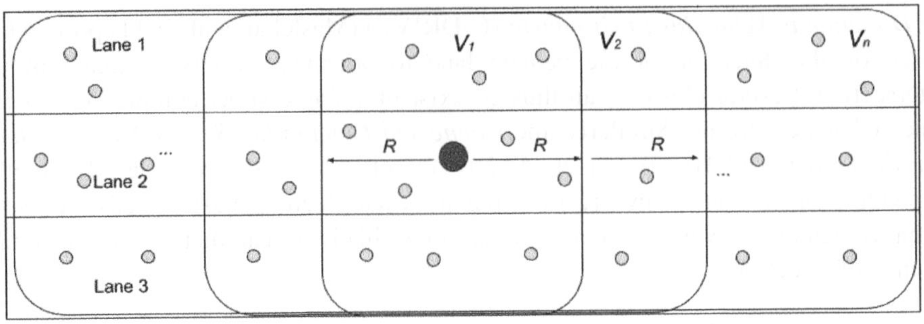

Figure 2.18. *Neighborhoods within n hops around a central node (in red) on a three-lane road. For a color version of this figure, see www.iste.co.uk/wahl/design.zip*

Density is used in several protocols to adjust operating parameters, of which:

– The transmission frequency of HELLO messages: by decreasing it when the density increases to avoid increasing the wait for accessing the medium or a saturation of the bandwidth, or by increasing it when the density decreases to promote more frequent detections of neighbors. This approach is adopted in particular in the *Connectivity Aware Routing* (CAR) (Naumov and Gross 2007) and *Density Based Clustering* (DBC) protocols (Kuklinski and Wolny 2009).

– The transmission power: by increasing it when the density is low to improve the connection between nodes that are probably more dispersed; or on the contrary by decreasing it when the density is high to reduce the range and virtually reduce the size of the direct neighborhoods, and thus the competition to access the medium (Figure 2.19). The *Dynamic Time-Stable Geocast* (DTSG) (Rahbar et al. 2010) protocol applies this method.

– Relay selection: several protocols use density as a criterion for the selection of relays, in particular in *Road Aware QoS Routing in VANETs* (RAQV) (Amjad and Song 2015), of cluster head in *Relatively Stable Clustering for Unbiased Environments* (ReSCUE) (Avcil and Soyturk 2015) and *Stable Clustering Scheme* (SCS) protocols (Arkian et al. 2015), in association with the position and speed of the nodes. Other protocols rather combine it with the particular configuration of the road, such as the *Landmark Overlays for urban Vehicular Routing Environments* (LOUVRE) (Lee et al. 2008) protocol, which gives priority to the selection of nodes located in the densest arteries relative to an intersection, or the *Geographic Stateless VANET Routing* protocol (GeoSVR) (Xiang et al. 2013), which systematically selects relays in multi-lane arteries, which are intuitively more frequented.

The direction of travel of vehicles with respect to the road is also a determining factor in the organization of communications. It can enable a protocol to detect neighboring vehicles traveling in the same or opposite direction to improve the estimation of communication time with neighboring vehicles or to detect the downstream or upstream of road traffic. The direction of a vehicle can be transmitted directly from the vehicle through the HELLO messages (Maslekar et al. 2011; Rawashdeh and Mahmud 2012; Luo et al. 2010; Song et al. 2010; Hadded et al. 2017), or it can be inferred from its successive positions as in the *Vehicular clustering based on the Weighted Clustering Algorithm* (VWCA) (Daeinabi et al. 2011) approach. In either case, an angular threshold can be defined to determine a common or opposite flow direction, in particular 45° in the *Angle-based Clustering Algorithm* (ACA) (Hadded et al. 2017) and *Mobility Prediction-based Directed Broadcasting* (MPDB) protocols (Cha et al. 2016).

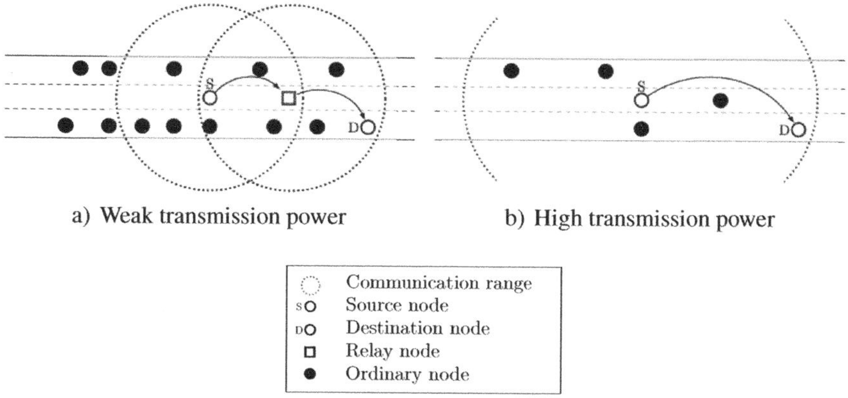

Figure 2.19. *Power variation according to the average number of neighbors (Rivoirard 2018)*

For the same absolute value of vehicle speed, the potential communication time will be longer between two vehicles traveling in the same direction than that between two vehicles traveling in opposite directions. Several protocols use the "travel direction" information of vehicles to estimate whether or not to include these vehicles in a routing strategy, in particular:

– When forming groups, the *clustering* protocols proposed in some previous studies (Luo et al. 2010; Song et al. 2010; Daeinabi et al. 2011; Maslekar et al. 2011; Rawashdeh and Mahmud 2012; Cha et al. 2016; Hadded et al. 2017) use this information to form clusters of vehicles that remain within communication range for as long as possible. Among the strategies adopted, the *Based on Euclidean distance clustering* (BEDC) (Tian et al. 2010) and *Vehicular Multi-hop Algorithm for Stable Clustering* (VMaSC) protocols (Ucar et al. 2013) form disjoint clusters for each flow direction, while *Particle Swarm Optimization Cluster* (PSO-C) (Harrabi et al. 2016) allows messages to transit between clusters with opposite traffic directions when this contributes to increase connectedness. The *Advanced Greedy forwarding* (AGF) (Naumov et al. 2006) and CAR (Naumov and Gross 2007) protocols utilize the direction information from neighboring vehicles to exclude from any election of nodes whose trajectory will leave the range area.

– During routing operations, the GPSR+PRedict protocol (Houssaini et al. 2017) proposed an enhancement to GPSR by using the flow direction of nodes to predict the future position of neighboring nodes and limit disconnections. This routing algorithm searches for an alternative to a route containing two nodes at the communication range boundary that are traveling in opposite directions. The idea is reused in the *Vehicle Assisted Data Delivery* (VADD) protocol (Zhao and Cao 2008) that operates by way of two relay election modes. On the one hand, there is the "current section" mode (*straightway*) that performs a conventional selection using a greedy algorithm. On the other hand, the intersection mode that selects the next relay, either by a so-called L-VADD approach, which aims to minimize the distance to the destination node (thus, Figure 2.20, the source node S would choose the relay node A, closer to the destinationnode D), or by the so-called D-VADD approach that prioritizes the direction of the nodes (thus, Figure 2.20 S would choose the relay node B which follows the same direction as D), or finally by a hybrid approach called H-VADD in order to avoid loops potentially induced by L-VADD.

– When transmitting directions of a particular area of road traffic. The *Relative Position Based Message Dissemination* (RPB-MD) (Liu and Chigan 2012), BBR (Wu et al. 2013a) and BBBR (Wu et al. 2013b) protocols make use of the direction of travel for selecting specific relays that allow packets to be transmitted only upstream or downstream of traffic.

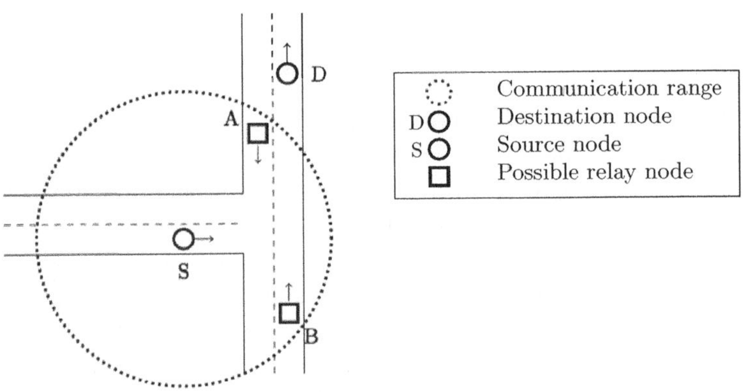

Figure 2.20. *Utilization example the VADD protocol in an intersection (Mukunthan 2015)*

The development of navigation aid applications makes it possible for vehicles to exchange information on their route, in particular when considering cooperative applications for the fluidity of road traffic. Vehicle route information is used by Souza et al. (2010) and Wang et al. (2008) to select cluster-heads for maximizing a travel time metric, especially in combination with the speed deviation information in the PBC protocol (Wang et al. 2008). The knowledge of the final destination is used to elect cluster-heads that ensuring cluster stability (Morales et al. 2011; Sethi and Chand 2017), in particular by choosing those with the furthest final destination or the highest remaining distance to travel as in the DBR protocol (Sethi and Chand 2017).

2.4.3. *Routing policies and quality of service metrics*

So far, all aspects of the organization of the routing protocols proposed have focused on how these protocols strive to minimize the impact of their own operation on communications resources in order to preserve them in favor of applications. Nonetheless, some applications have specific requirements, such as in terms of bandwidth, packet delay or loss. Consequently, for a routing protocol to take these requirements into account, additional efforts are necessary in addition to its own initial organization in order to enable these applications to effectively utilize them. This point, generally identified by the notion of quality of service (QoS) routing, is crucial for the development of applications based on ad hoc routing protocols and largely explains the low number of effective solutions currently deployed on this model, especially in the road environment. In particular, QoS protocols in VANETs must take into account the physical constraints of communication resources in terms of bandwidth, communication range and message loss due to both the nature of the medium and the protocol accessing the medium. Frikha (2010) made a distinction between four families of metrics for the algorithms of QoS-routing protocols dedicated to VANETs: those related to network topology, link quality, traffic load and link mobility (Table 2.1).

Regarding its measurement (Sarr 2007), a metric can be locally measured because it is directly accessible by the node (number of direct neighbors, for example, by consulting the local table of neighbors within one hop) or externally to the node through passive observation (by way of sent or received control messages, for example, by counting them). It can be also actively measured (by creating control messages specifically dedicated to calculating the value of a metric), or still indirectly (by inserting into existing the control messages data for the computation of a metric value). Depending on the computation method, a metric is said to be additive, multiplicative, concave or convex. Let a path R comprising n successive links $(E_1, E_2, ..., E_n)$, Wang and Crowcroft (1996) identified the following rules, for any metric M, enabling the calculation of its overall measure on the path under consideration from its measurements on the links taken individually:

– additive: $M(R) = \sum_{i=1}^{i=n} M(E_i)$;

– multiplicative: $M(R) = \prod_{i=1}^{i=n} M(E_i)$;

– concave: $M(R) = \min\{M(E_1), M(E_2), ..., M(E_n)\}$;

– convex: $M(R) = \max\{M(E_1), M(E_2), ..., M(E_n)\}$.

Metric	Optimization			Classification				Mathematical Properties					Measurement method				
	Delay	Losses	Throughput	Topology	Link quality	Mobility	Traffic load	Additive	Multi	Concave	Convex	Objective	Internal	External	Passive	Active	Hybrid
ETT	✓				✓			✓				↘		✓		✓	✓
NoH	✓			✓				✓				↘	✓				
ETX		✓			✓		✓			✓		↘		✓	✓	✓	✓
LET		✓			✓	✓				✓		↗		✓			✓
BW			✓				✓			✓		↗		✓	✓	✓	✓
TP			✓		✓					✓		↗		✓	✓	✓	✓
AH			✓		✓					✓		↗	✓		✓	✓	✓

Table 2.1. *Quality of service metrics*

Some of the most referenced metrics in studies of ad hoc routing protocols include:

– The delay, usually in the form of two different metrics: on the one hand, the *End-to-End Delay* (ETT), which is the time that a packet takes to reach its destination from the time it is created by the protocol associated with the source node until it is received by the protocol corresponding to the destination node level. This additive metric accumulates the time spent by the packet in the queues of the different layers of the OSI model, the time for processing it, the time for its transmission(s) and its propagation(s) between nodes. It is used by the *Adaptive QoS for Vehicular safety services* (AQVA) protocol (Kalivaradhan and Thangavelu 2010), the *QoS Bee Routing* (QoSBee) *Protocol* (Bitam and Mellouk 2011), *Intersection based Routing with QoS support in VANET* (IRQV) (Li et al. 2015) or the *VANETs' Quality of Service by Adaptive* (VOA) *protocol* (Yan et al. 2012) when electing relay nodes in their reactive route search procedure. It is also used as a cost metric for finding a route in greedy mode in the *Backbone Assisted Hop Greedy* (BAHG) protocol (Sahu et al. 2013). On the other hand, the hop count (NoH)-related metric counts the number of relay nodes between the source node and the destination node. Due to the fact that nodal processing times are incompressible and typically longer than propagation delays, this metric can be thought of as an indirect measure of the end-to-end delay for a path, which also increases with the number of relay nodes in a path. By minimizing this metric, the RAQV (Amjad and Song 2015) and *Reliable QoS Routing for VANETs based on AODV* (RQ-AODV) protocols (Zheng et al. 2014) search for paths that will reduce packet delays.

– The packet loss rate, especially in the form of two different metrics: on the one hand, the *Expected Transmission Count* (ETX) corresponding to the number of transmissions required to send a packet between two nodes, including retransmissions due to poor reception. It is often indirectly computed using HELLO packets. Since the latter are periodically sent, their large number makes it possible to perform statistical measurements of the ETX metric over time. The computation of ETX by way of HELLO packets is spurious, because these packets are smaller than application packets, and are therefore less likely to be lost (Yélémou 2012). However, it makes it possible to approximate the value of the ETX metric without sending an additional packet, and it is quite possible to include certain application packages in the statistic as long as their periodicity and size are well defined. On the other hand, link expiration time (LET) measures the time period in which two nodes are within direct communication range of each other. It is calculated by taking into account node mobility and is used by a large number of reactive protocols in the election of the next relay node (for instance, in *Adaptive Routing Protocol Based on QoS and Vehicular Density in Urban VANETs* (ARP-QD) (Sun et al. 2015), RQ-AODV (Zheng et al. 2014), *Optimal Next Hop Selection* (OHNS) (Lin and Huang 2012), VOA (Yan et al. 2012), *Mobility-Based Scheme* (MBS) (Ren et al. 2017) and RAQV (Amjad and Song 2015)). The cumulative loss rate in a path is computed as a multiplicative metric. Nevertheless, the diversity of routes between a given source-destination pair logically decreases the estimated loss rate associated with all routes, as shown in particular by Gantsou et al. (2009).

– The link quality, which refers mainly to three metrics: first, the available *Bandwidth* (BW), which counts the number of times the communication medium is free compared to the number of times it is busy. It is particularly used by the OHNS (Lin and Huang 2012) and AQVA reactive protocols (Kalivaradhan and Thangavelu 2010) as a criterion for selecting intermediate nodes in a route. It is also used by the *Dynamic Backbone-Assisted Medium Access Control* (DBA-MAC) protocol (Di Felice et al. 2011) to elect cluster-heads. The *QoS-Optimized Link State Routing protocol* (QOLSR) (Ge et al. 2003) (with its variants proposed by Badis and Al Agha (2004), Badis and Al Agha (2005), Leguay et al. (2006), Nguyen and Minet (2006) and Sondi et al. (2010)) and the QoS-OLSR protocol (Wahab et al. 2013) make use of the bandwidth metric to improve the choice of MPR nodes. Next, the *Transmission Power* (TP) used in the DBA-MAC protocols (Di Felice et al. 2011) and *On-Demand Virtual Backbone BroadCAST Protocol* (OVB-CAST) (Shi et al. 2009) to elect the nodes proposing the best radio links as cluster-heads. The VWCA protocol (Daeinabi et al. 2011) implements the *Adaptive Allocation of Transmission Range* (AATR) algorithm, which can be used to vary the transmission power based on the number of neighboring nodes. Starting from a low power at node initialization, the AATR algorithm increases the value of the transmission power until the node obtains a number of direct neighbors deemed satisfactory. Finally, the *Antenna Height* (AH) is also an indicator of the quality of the radio link. The larger the antenna size (typically that of a bus or truck), the greater the communication range and reception quality. This metric is considered in the BBR (Wu et al. 2013a), BBBR (Wu et al. 2013b) and BBNC protocols in the node selection function of their *backbone*.

A large number of studies propose approaches combining several QoS metrics both in route retrieval and in the election of cluster-heads. Sondi et al. (2013) provided a formal framework that includes a description of an ad hoc network from the perspective of multi-criteria QoS routing, with metric composition rules, the formulation of cluster-head selection and path computation problems, as well as design guides for adapted algorithms. This work applied to the OLSR protocol constitutes an entry point to explore, through the previous work referenced in it and the recent work that cites it, the topic of quality of service in routing protocols for ad hoc networks. One of the latest reviews of the literature specifically dedicated to quality of service in ad hoc vehicle networks is available in Belamri et al. (2021).

2.5. Functional design approach: concepts and case study

2.5.1. *Idea and approach*

The development of a communication system based solely on V2V communications still incurs many design challenges despite the proliferation of propositions in the literature (Katiyar et al. 2020). The design approaches generally used in this work include the functional approach, which is the most represented. It consists of developing the routing protocol by adopting choices and functionalities whose assembly and operation make it possible to respond to all of the requirements of the vehicle network.

The previous sections of this chapter have described the basic concepts as well as the various functionalities observed in routing protocols for ad hoc networks in general and vehicle networks in particular. Through a case study, this section aims to illustrate how the functional design approach can be implemented at all design stages for a routing protocol for VANETs. The purpose of the case study routing protocol is to enable the implementation of a virtual infrastructure based solely on vehicles for managing communications induced in particular by cooperative applications deployed in vehicles.

2.5.1.1. *Case study protocol positioning and orientation*

Historically, the organizational models proposed in routing protocols for MANETs have been explored for VANETs. In addition to conventional (*blind flooding*), whose inefficiency is established in many respects in the vast majority of studies, the literature distinguishes organizational models on the one hand, which do not maintain topology and generally operate on demand and, on the other hand, those that structure the network and thus have the ability to proactively operate.

The first group include reactive protocols such as *Dynamic Source Routing* (DSR) (Johnson et al. 2007) and *Ad-hoc On Demand Distance Vector* (AODV) (Perkins et al. 2003), as well as all protocols derived from them for application in VANETs. We did not adopt this approach, and more generally the organizational models that derive therefrom, because their mode of operation is precisely based on the idea that the network should not be structured. Admittedly, each node in the network participates in the retransmission of messages allowing the other nodes to find paths to the intended destinations, as well as in routing traffic to the destinations. However, the fact that nodes do not maintain a topology and that it is not possible to control the periodicity of message transmissions makes their implementation and maintaining a virtual infrastructure based on this approach difficult.

The second group refers to proactive protocols such as DSDV (Perkins and Bhagwat 1994) and OLSR (Clausen et al. 2003) and hybrids, as well as all of the variations that have stemmed therefrom, especially for VANETs. In a proactive approach, each node maintains a routing table in addition to its one-hop neighbor table and periodically broadcasts routing packets across the network, independently of any application traffic. These routing packets contain the list of routes known to the node and enable routing tables to be managed. Hybrid approaches combine the proactive approach for the near neighborhood up to a certain number of hops in the topology and a reactive approach for destinations outside that area. The organizational models derived from these approaches result in the construction of a virtual infrastructure. These include models leading to the formation of a tree structure, *clustering* methods leading to the formation of a cluster of nodes represented by one or more particular nodes (cluster-head) and methods leading to a mesh where each node selects a set of nodes (the multipoint relays) enabling it to reach its neighbors further than one hop away. They will then be considered for the case study.

Two new approaches have been developed specifically for ad hoc vehicle networks. In the first approach, known as opportunistic (Chakchouk 2015) and from which the proposition (Li 2015) for VANETs is derived, the network nodes make the routing decision according to the state of the network and on the basis of predefined characteristics. Although its main purpose is not building a structure, it nevertheless makes it possible to achieve a certain rate of connectedness on each road segment. Nevertheless, in order to support cooperative applications such as ego-localization that requires the formation of disjointed groups of vehicles, group formation strategies should be added to the opportunistic approach. This approach may therefore be of interest for our goal of achieving a virtual infrastructure. The second approach combines all of the so-called geographical protocols (Boussoufa-Lahlah et al. 2018), including GRP (Zhiyuan 2009). The implementation of the localization service is not clearly defined in the GRP specification. Since our notion of virtual infrastructure was designed for the purpose of operating when no terrestrial infrastructure is available, we will only retain the possibility that each node knows its own position from a geostationary positioning system. In fact, the notion of a white zone is anecdotal with regard to this type of system and its functioning is almost never disturbed by human activity on Earth. Therefore, it is not part of the infrastructures that we exclude from our approach.

2.5.1.2. *Spatial, functional and application constraints*

In contrast to the organizational models proposed for MANETs, which in general assume free mobility in an open space, VANETs are spatially restricted by the configuration of traffic lanes. As such, if the basis of the study is reduced to any segment between two intersections, the VANET evolves within a longitudinal space comprising one to several traffic lanes and one to two directions. If a communication infrastructure had been deployed to cover this space, an access point (*Road Side Unit* (RSU)) would have been deployed in the center of a portion of the road segment, approximately equal in length to the range of the communication technology. For virtual infrastructures, since the access point is also a vehicle, one of the vehicles found at that time in the segment therefore simply has to operate as an access point over a distance equivalent to a communication range. Intuitively, we assume that the same group of vehicles travels through two successive segments S_1 and S_2. When a physical infrastructure is involved, the segments and their infrastructure elements are fixed. These are the vehicles that successively attach to the various access points of S_1 until they exit the segment, and then do the same thing on S_2. For vehicles traveling in the opposite direction, it is exactly the same thing, except that they move from S_2 to S_1. When considering a virtual infrastructure, the access points are vehicles spaced apart from one another by a length close to the range. The correct picture consists to consider, assuming that speed is homogeneous, that the cluster of vehicles forming the virtual infrastructure is stationary and that segment S_1 slides below the cluster of vehicles before giving way to S_2. This illusion does not work in the presence of a second traffic direction, because it is not possible to make the space slide in two opposite directions at the same time. Consequently, for a virtual infrastructure, a cluster of access point vehicles is required for each traffic direction on the segment under consideration.

In order for the proposed virtual infrastructure to comply with the European standard for vehicular communications (ETSI 2013), it should make it possible to implement the four communication scenarios specified for vehicle networks (Figure 2.21), namely, point-to-point, point-to-multipoint, *GeoAnyCast* and *GeoBroadcast*.

In terms of application constraints, in order to be able to navigate within a flow of cars on a motorway or departmental environment, each vehicle will be able to periodically share local information with its immediate neighbors (*Cooperative Awareness Message* (CAM) (ETSI 2019a)) in order to synchronize its movements in accordance with those of its neighbors. They must also be informed as soon as possible of any change in the flow of vehicles (such as stopping, slowing down, speeding up traffic and obstacle avoidance) in order to modify their trajectory in advance in collaboration with those of vehicles in their vicinity. This second constraint requires the ability to transfer aperiodic messages (*Decentralized Environmental Notification Message* (DENM) (ETSI 2019b)) over distances greater than the one-hop neighborhood, between a network node and one or more destination nodes. In addition, in order to enable the optimal operation of cooperative applications such as ego-localization and extended perception, the network nodes must form disjointed clusters. This partition can be achieved by imposing that each vehicle in the network be associated with a single vehicle acting as an access point node in the virtual infrastructure.

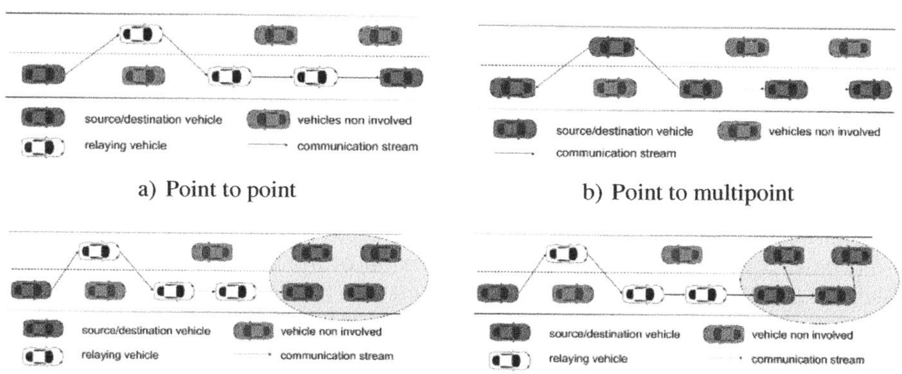

a) Point to point

b) Point to multipoint

c) *GeoAnycast*

d) *GeoBroadcast*

Figure 2.21. *Communication modes supported by the GeoNetworking protocol (source: ETSI 2013). For a color version of this figure, see www.iste.co.uk/wahl/design.zip*

2.5.2. *Overview of the case study protocol:* Chain-Branch-Leaf

A functional design approach generally relies on a geometric intuition to propose an organization that makes it possible to best implement the targeted functionalities. The longitudinal configuration of roads in vehicular networks strongly conditions the type of organization considered in this case study. Based on the considerations mentioned in section 2.5.1, we have synthesized a proposition (Rivoirard 2018), which can be categorized as one-hop *clustering* methods with backbone formation, but with the originality of explicitly responding to all of the functional requirements mentioned. This proposition, which is the subject of this case study, is called *Chain Branch Leaf* (CBL). It mainly focuses on creating clusters of one-hop communicating nodes with a single hierarchical level, thus forming a hierarchical topology. Members of a cluster are within direct range of communication with an elected cluster-head. Therefore, the size of the clusters is de facto limited to twice the radio transmission range. The cluster-heads form a dominant connected set, in this case the virtual infrastructure, in the flow direction being considered, connecting disjointed clusters (section 2.5.1.2).

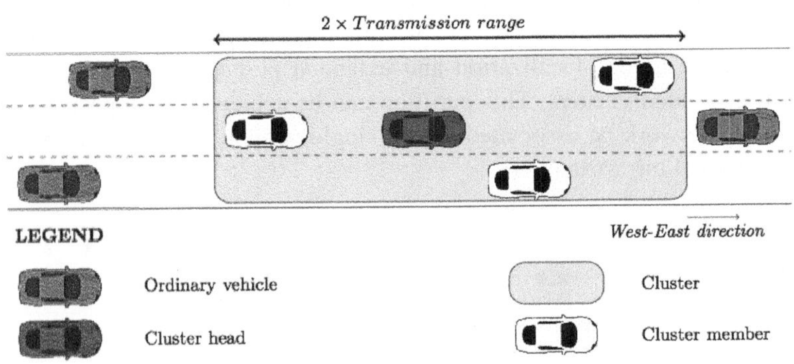

Figure 2.22. *Illustration on a three-lane highway of a cluster of vehicles created around a cluster-head. The size of the cluster is limited to the neighborhood within one hop of the head, that is, twice the transmission range of the onboard radio technology (Rivoirard 2018). For a color version of this figure, see www.iste.co.uk/wahl/design.zip*

2.5.2.1. *Concept of branch and leaf nodes*

The organization of CBL is based on the specification of two types of nodes between which each network node oscillates continuously, according to rules locally applied by all nodes in its vicinity:

– A branch node is a node responsible for a cluster of leaf nodes, thus operating as a cluster-head. It is chosen by other nodes (branch or leaf) to work as a relay to the rest of the network. As with most proactive protocols, a branch node performs neighborhood discovery through periodic HELLO messages and broadcasts its routing tables throughout the network, in particular by means of TC messages. It contributes to building a chain of relay vehicles that travel in the same direction on the road, thus materializing the virtual infrastructure. In the chain it is a part of, it ensures that it maintains a valid link with an upstream branch node and a downstream branch node in its flow direction and elects an upstream branch node according to predefined criteria. In addition to being a potential source of application traffic, a branch node also ensures the relay for retransmitting messages between network nodes that are more than one hop away.

– A leaf node is an ordinary node that has no special responsibilities and that performs neighborhood discovery by means of periodic HELLO messages. It is attached to the branch node of the portion of road in which it operates. When none are detected, it chooses as a branch node, according to well-defined criteria, one of the neighboring leaf nodes circulating in the same direction.

It should be noted that the specification of branch and leaf nodes creates disjoint clusters as well as links of different kinds. Because each node is a member of a single group identified by the cluster-head (branch node), the sets of leaf nodes from two adjacent clusters in the same chain form disjoint clusters. From the notions of branch and leaf nodes, two types of links can be inferred: those between the leaf nodes of a cluster and their branch node or cluster-head (intra-*cluster*), as well as pairwise links between branch nodes that define chain links (inter-*cluster*).

2.5.2.2. *Notion of chain*

The construction of chains is ensured by branch nodes. A chain is a virtual backbone network composed of an ordered sequence of branch nodes that flow in the same direction. Therefore, a chain is the virtual infrastructure created by each flow direction and in which each branch node is one of the links. With respect to routing, chains support the routing of the traffic of topology (TC) control messages retransmitted from branch node to branch node. For application messages, the chain provides a service for transferring messages from a source node to one or more destination nodes located more than one hop apart. Figure 2.23 shows how a chain is formed on a three-lane highway. It should be noted that the set of branch nodes forms a dominating set connected in its related component (the chain). This is because any node in the network is either a branch node or a leaf node directly attached to a branch node, and any branch node can reach any node in the same chain by a route whose intermediate nodes are branch nodes.

2.5.2.3. *Specification of the propagation direction of an application message*

To ensure that application traffic can be transferred upstream or downstream in a chain, or even broadcast to all nodes within the chain neighborhood, a field containing direction information is added to the header of application packets. Let "Dir" be the name of this header field that encodes the instructions such that one or all of the following instructions can be executed for a given packet: transmit downstream of the traffic, transmit upstream of the traffic, transmit to near neighborhood, and forward to all nodes in the network.

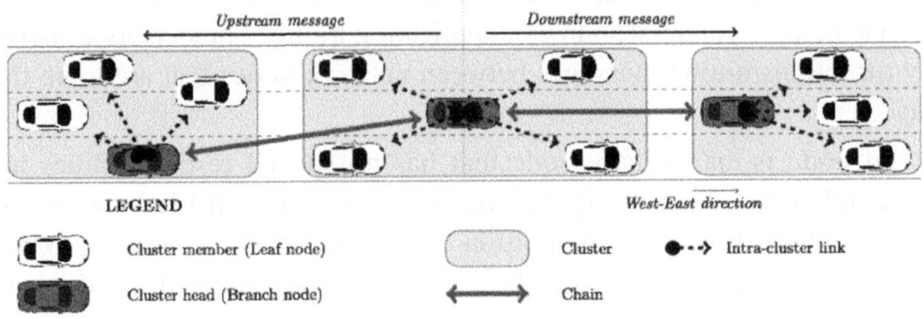

Figure 2.23. *Illustration of the creation of a chain in a motorway scenario exemplifying the concept of sending messages according to the flow direction, downstream or upstream of the traffic (Rivoirard 2018). For a color version of this figure, see www.iste.co.uk/wahl/design.zip*

Through this field, the algorithm controls the broadcast of an application message to limit the risk of flooding in the VANET (*broadcast storm*) that could result thereof. For example, depending on the encoding of "Dir" in the application message header, it will be possible to send an alert message (*Decentralized Environmental Notification Message* (DENM)) about the risk of downstream traffic jams, or an alert message about the presence of a priority vehicle (police, firefighter, etc.) upstream, or a periodic message to signal its presence to neighboring vehicles (*Cooperative Awareness Message* (CAM)).

2.5.2.4. *Definition of the structuring metrics*

Once the overall vision of the organization to be implemented has been established, it is necessary to analyze vehicle dynamics in order to identify the parameters that will be decisive in the implementation and maintenance of this organization. In this case study, the creation of the CBL chain requires the use of the following four decision metrics attached to vehicles:

– The position $[X_i, Y_i]$ of any vehicle N_i that is used by CBL to compute the distances between vehicles. As shown in section 2.5.1, this position information is obtained by the vehicle via a geostationary positioning system.

– Vehicle speed, which includes the absolute speed V_i and the value of the steering angle σ_i averaged over one second. The absolute speed is used for ordering the vehicles and the steering angle is used for classifying them according to their direction of travel. Their respective values (V_i, σ_i) can be obtained by means of an inertial system and an angle sensor, or can be deduced by tracking successive vehicle positions.

– The relative degree (*Degree*) of a neighboring node N_j one hop away from N_i is the number of symmetric neighbors of that node N_j after excluding node N_i and the nodes in the neighborhood within one hop of N_i. This metric is used to identify one-hop neighbor nodes that provide knowledge about new neighbors.

– *Connection time* (CT/Ctime) is computed from the position, speed and direction of the knots. CT is used by a leaf node to choose its branch node. It is defined as the time during which two nodes could communicate if they maintained a constant speed. We took the metric defined in Li et al. (2013) and Jemaa (2014) as connection time and calculated using equation [2.3]:

$$CT(N_i, N_j) = \frac{-(ab+cd) + \sqrt{(a^2+c^2) * P^2 - (ab-bc)^2}}{a^2+c^2} \qquad [2.3]$$

$$\left\{ \begin{array}{l|l} a = V_i cos(\sigma_i) - V_j cos(\sigma_j) & b = X_i - X_j \\ c = V_i sin(\sigma_i) - V_j sin(\sigma_j) & d = \ Y_i - Y_j \end{array} \right.$$

with P as the maximum radio range of the network technology employed.

Figure 2.24 shows the connection time between two vehicles according to their inter-distance and relative speed when the network technology used is the *Institute of Electrical and Electronics Engineers* (IEEE) 802.11p. Figure 2.24(b) shows the results obtained for two vehicles traveling in the same direction, and Figure 2.24(a) shows a scenario with two vehicles traveling in opposite directions. In this example, we observe that the maximal values taken by the CT for two vehicles traveling in opposite directions are comprised between 5 and 20 s, while for two vehicles going in the same direction of travel, this connection time can exceed 100 s.

Sometimes, the parameters set may not be enough to fully determine the evolution of the targeted organization. In the case study, we had to introduce *ChainTime* as a new metric for the choice of upstream and downstream branch nodes by a given branch node. It is computed based on the calculation of the CT and is defined, for two branch nodes in a chain, as the minimum time between their *Connection Time* (CT) and the time elapsed before one of the two nodes overtakes the other. *ChainTime* is given by equation [2.4]:

$$ChainTime(N_i, N_j) = \begin{cases} CT(N_i, N_j) if\ N_i\ downstream\ of\ N_j\ and\ V_j < V_i \\ CT(N_i, N_j) if\ N_i\ upstream\ of\ N_j\ and\ V_j > V_i \\ min \left[CT(N_i, N_j), \frac{\sqrt{(X_i - X_j)^2 + (Y_i - Y_j)^2}}{|V_i - V_j|} \right] otherwise \end{cases} \qquad [2.4]$$

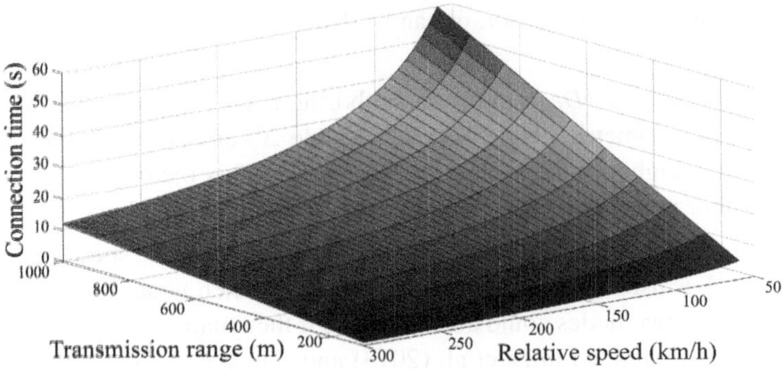

a) Example of two vehicles traveling in opposite directions

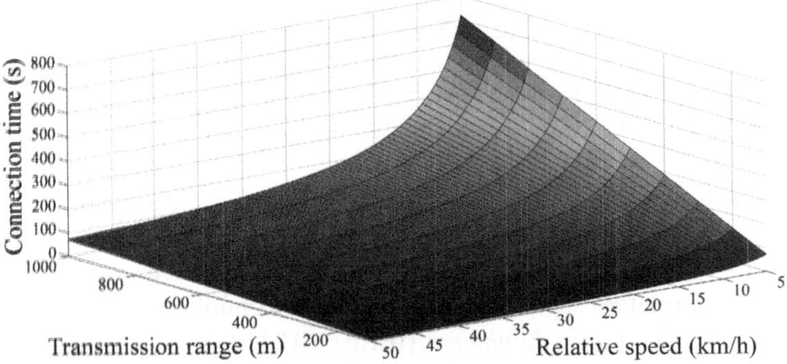

b) Example of two vehicles traveling in the same direction

Figure 2.24. *Connection time (CT) depending on relative speed and transmission range (Rivoirard 2018). For a color version of this figure, see www.iste.co.uk/wahl/design.zip*

An immediate effect of the *ChainTime* parameter is depicted in Figure 2.25. The average time taken by a leaf node to attach to the same branch node increases (Figure 2.25(b)), almost from simple (CBL_v1) to twice (CBL_v2), even though overall, this adaptation reduces the amount of time a node keeps the status of branch node (Figure 2.25(a)). The interpretation is that a node remains a branch for an average time period, which is roughly the same as the time during which any leaf node remains attached to the same branch node. This reflects a certain stability in *clusters*. As a result, the traffic generated by any node takes longer to travel through the same relay node, which is often the case in an area where fixed infrastructure is deployed. This underlines the fact that we choose to give preference to the connection time between leaf nodes and their branch node rather than to the amount of time a node remains a branch, regardless of its relationship to its leaf nodes.

Therefore, it is clearly understandable that sometimes the parameters will influence very positively in favor of a certain functionality and less in favor of another. Under these conditions, the designer will still have to make the decision, and devise an approach or impose conditions that allow us to shift from one approach to another as the network evolves. A detailed presentation of the CBL implementation, including the formats of the messages, the encoding of variables and their use in the management of the virtual infrastructure, as well as the parametrization of the time threshold, can be found in Rivoirard et al. (2020).

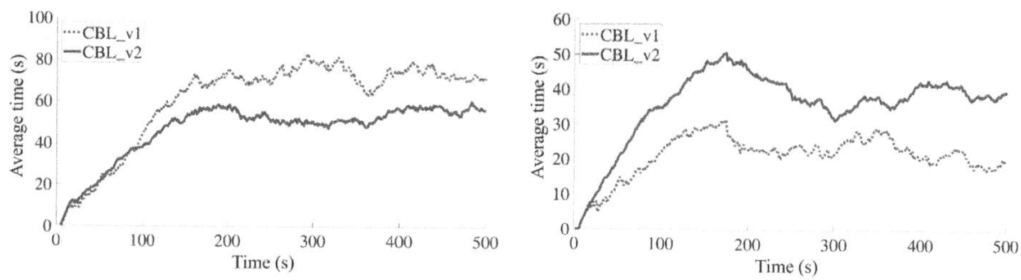

Figure 2.25. *Increase of the stability of CBL by means of the ChainTime parameter: CBL_v1 is a version of CBL without the ChainTime parameter, CBL_v2 is a version with ChainTime. For a color version of this figure, see www.iste.co.uk/wahl/design.zip*

2.5.2.5. *Construction algorithms for the CBL structure*

A detailed presentation of the algorithms involved in the implementation and maintenance of the virtual infrastructure using CBL is provided in Rivoirard et al. (2018). Improvements that allow us to extend the attachment time of leaf nodes to their branch, thus enhancing the overall stability of the virtual infrastructure, can be found in Wahl et al. (2021). The main objective sought for in this case study is to describe the ideas that guided the implementation of the proposed mechanisms to achieve the functional objectives set for the routing protocol.

2.5.2.5.1. Transmission process for a HELLO message

For a constant time interval *HELLO_INTERVAL*:

1) a HELLO message is created by a node N_i;

2) it is placed in the queue for the next transmission;

3) N_i remembers the creation date of the message in its local variable T5.

The HELLO message is created with the extended format by CBL and contains the values of the local variables of the node N_i described in Rivoirard et al. (2020).

These values are regularly updated by N_i in the interval *HELLO_INTERVAL* between two HELLO transmissions; and are updated each time a HELLO message is received from a neighboring node.

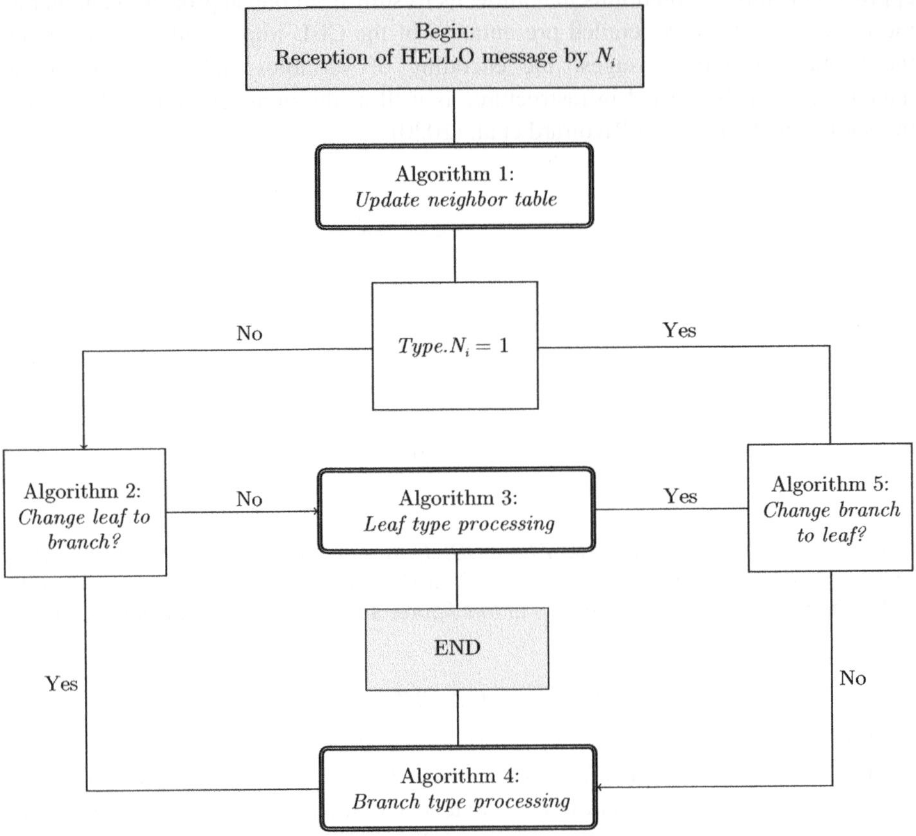

Figure 2.26. *Procedure upon receiving a HELLO message (Rivoirard 2018)*

2.5.2.5.2. CBL block diagram when receiving a HELLO message

When the network is initialized, no group is formed, all nodes in the network are leaf nodes. The HELLO messages exchanged by the nodes trigger the mechanism for creating the hierarchical topology. When a HELLO message is received from a neighboring node, any node N_i performs the procedure shown in Figure 2.26. In the first task, Algorithm 1 updates the routing table containing the list of one-hop neighboring nodes. Then, depending on the type (branch or leaf) of the receiving node N_i, the tasks executed will be those of Algorithm 5 (if N_i is a branch node) or Algorithm 2 (if N_i is a leaf node). The function of Algorithm 5 is to reset the branch node N_i to a leaf node. The function of Algorithm 2 is to transform the leaf node N_i into a branch node if the sender node of the received message has elected it as relay node. When Algorithms 2 and 5 terminate, depending on the resulting nature of the receiving node N_i, that is, leaf or branch, Algorithm 3 (if N_i is a leaf node) or 4 (if N_i is a branch node) is executed. Algorithm 3 enables the leaf node N_i to perform, verify, or maintain the election of its branch node. Algorithm 4 enables the branch node N_i to carry out, verify or maintain the election of branch nodes in its chain, upstream and downstream of the direction of road traffic.

This simple operation makes it possible for each node, between two successive transmissions of its HELLO messages (*HELLO_INTERVAL*), to update its status based on its knowledge of its situation and that of its neighbors obtained from the HELLO messages received from them during this period of time. Therefore, it can decide its status and its possible branch node choices (or front and rear branch nodes) and notify its direct neighbors when its next HELLO message is sent. This operation mode allows maintenance of the *clusters* and the backbone (the virtual infrastructure) based on a single message type and time parameter. Actually, it is possible to control the evolution of the structure somewhat finely by adjusting the parameter *HELLO_INTERVAL* according to the speed of road traffic.

2.5.2.5.3. Some properties of CBL

Since it is not possible to examine how the algorithm works in detail, a few elements about its operation should be pointed out.

– In the organization imposed by CBL, branch nodes operate as relays when routing information between nodes is more than one hop away. Each is the *cluster-head* of an empty set or comprising the leaf node(s) that elected it. Groups defined according to the CBL scheme therefore exhibit the following property that can be used for the implementation of cooperative applications: given that a leaf node selects only one branch node and that a branch node attaches to at most two branch nodes of the same chain (one upstream its position and the other downstream), each cluster of leaf nodes aggregated around a branch node defines a set that is disjoint of those built around neighboring branch nodes. This property of CBL will be fundamental for the implementation of cooperative applications.

– CBL employs seven weights (L1, L2, L3, C1, C2, C3 and α) of which the first four, the C1, C2, C3 and L3 thresholds, have been set to improve the algorithm responsiveness. The value of the other thresholds (L1 = 0.5, L2 = 5 and α = 0.5) was arbitrarily chosen. It should be noted here that it is usual for experts to set certain parameters of the algorithm arbitrarily based on their experience. Nevertheless, under these conditions, it is necessary to perform an analysis of the sensitivity of the algorithm to parameters whose value has been arbitrarily set (here the L1, L2 and α thresholds) to ensure that they do not have a greater influence than those whose adequate values have been formally established (here the C1, C2, C3 and L3 thresholds). This sensitivity analysis achieved on the metric "total number of leaf nodes in the network" shows that the algorithm is only slightly impacted. As a matter of fact, for constant threshold values of C1, C2, C3 and L3, the value of the metric varies only very slightly depending on the combinations of the admissible values of L1, L2 and α (Figure 2.27).

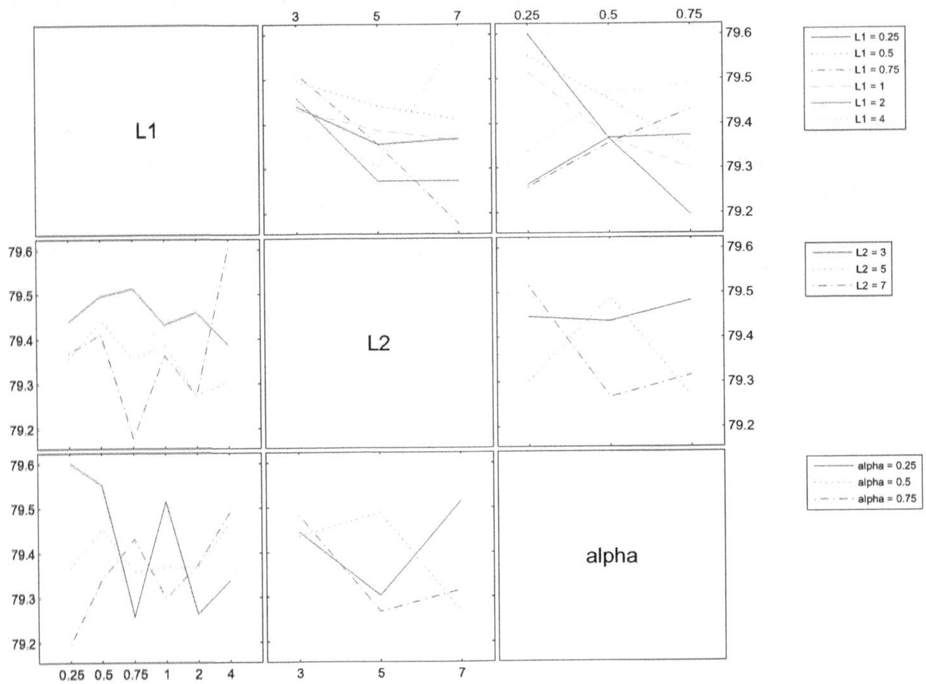

Figure 2.27. *Impact observed via Ms1 between t = 150–300 s of the weight coefficient L1, L2 and alpha on the CBL structure – scenario S5 case A (y-axis in %). For a color version of this figure, see www.iste.co.uk/wahl/design.zip*

The analysis showed that several situations could lead to the appearance of parallel chains in the same flow direction. The problem could be solved by various mechanisms detailed in Rivoirard (2018), in particular by restricting the permission to extend the chain only upstream of the direction of travel. By no longer allowing the chain to be extended simultaneously upstream and downstream, a cause for conflicting choices that could lead to parallel chains is avoided. The effect of this choice on the stability of the chains is shown in a comparison of the structures of the virtual infrastructure before and after this measurement (Wahl et al. 2021). On a two-way road, the organization imposed by CBL should result in the construction of two chains, ideally one in each direction of traffic. Figure 2.28 shows that prior to the introduction of the upstream extension restriction only, the appearance of parallel chains and other chain breaks could lead to the observation of up to eight parallel chains (CBL_v1). The time corresponding to the observation of two chains was limited to one-third (150 s) of the total duration of the scenario (500 s). After the restriction is applied, this duration is almost double the initial value (CBL_v2) and fewer and less frequent divisions occur in the chains.

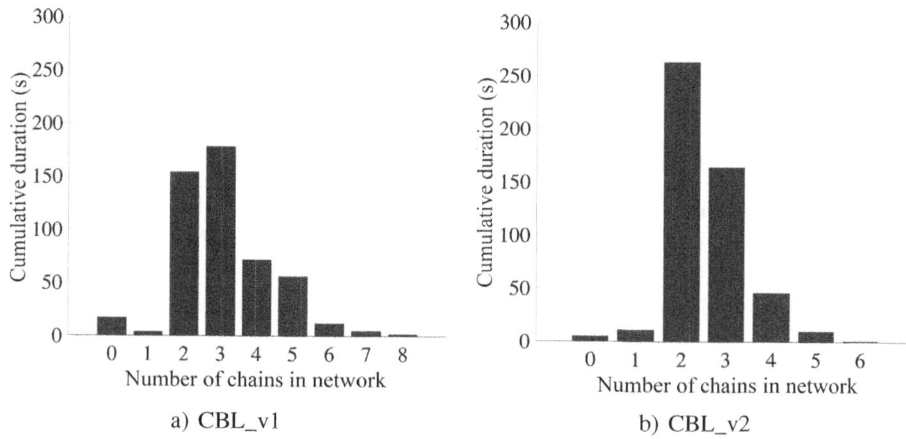

Figure 2.28. *Cumulative duration for each value of the number of chains observed*

2.5.3. Modeling CBL in OLSR: CBL-OLSR

A detailed presentation of the implementation of CBL in OLSR is available in Rivoirard et al. (2020). In this section, we reuse part of the presentation that was made in Rivouard's (2018) thesis.

Although CBL is a proactively operating organization model, it does not specify an entire routing protocol. For example, CBL does not specify route–search mechanisms. In order to be operational, it therefore needs to be implemented in a routing protocol. In addition to a geostationary positioning system that allows each node to know its own position, the operation of CBL is essentially based on the periodic transmission of messages. Any implementation of CBL in an existing routing protocol therefore requires, if there is not already an existing one through the applications deployed (CAM, for example), a definition for the periodic transmission of messages.

Among the well-known routing protocols, we have presented the OLSR protocol which is particularly well suited for high-density networks (Lebedev 2006). Another reason why OLSR is an interesting candidate for a CBL implementation has to do with its characteristics. A comparative evaluation of OLSRs integrating CBL with multiple protocols for MANETs, proposed in Wahl et al. (2021), confirms the relevance of this choice. OLSR is an ad hoc routing protocol whose operation is based, on the one hand, on the periodic transmission of HELLO messages also necessary for the operation of CBL and, on the other hand, on the definition of two types of nodes: the *Multipoint Relays* (MPRs) nodes (relay nodes of the protocol) and ordinary nodes such as for CBL with branch nodes and leaf nodes. The implementation of CBL within the OLSR protocol leads to the proposition of the *Chain Branch Leaf-Optimized Link State Routing protocol* (CBL-OLSR) variant. The information required for the operation of CBL will be added to the HELLO messages sent by OLSR.

2.5.3.1. *Mapping MPR to branch nodes*

The OLSR protocol performs a structuring of the network through the MPRs nodes in order to save radio resources. The selection principle of MPRs nodes is the following: each node N_i in the network selects a set of one-hop neighbors thereby giving it the possibility of communicating with its two-hop neighbor. MPR nodes are used as relays such as the node that elected them can communicate within its two-hop neighborhood. Only the nodes in this set retransmit messages sent in *broadcast* by their elector.

The CBL scheme makes it possible, as in the OLSR protocol, to form a hierarchy within the network by making a distinction between two types of nodes. The implementation of CBL within OLSR has therefore naturally moved toward the use of CBL branch nodes in replacement of MPRs nodes. The branch nodes then inherit the same functionality as the MPRs nodes. These are nodes that relay messages sent in *broadcast* mode (such as TC messages). It should be noted that the implementation of CBL within OLSR impacts the choice of routes initially designed by CBL. This is because the OLSR protocol performs a route search by the shortest path algorithm computed using the number of relay nodes in a route. It is possible for a route to contain non-MPRs nodes, whose function is then to relay a message sent in *unicast* mode using that route. Therefore, since the CBL implementation within OLSR proposed here consists of defining branch nodes as MPRs nodes and leaf nodes as non-MPRs nodes. It is possible that a leaf node relays a message sent in *unicast* mode.

2.5.3.2. *CBL-OLSR packet format and communication services*

The CBL implementation within the OLSR protocol requires a modification to include the direction information of "Dir" messages (section 2.5.2.3). This information is added to the OLSR message header at the message type level, which is initially encoded in 8 bits. Yet, in the initial version of the OLSR protocol, the message type can only take integer values comprised between 1 and 4.

The proposed implementation uses the first four bits to encode the "Dir" information (see Table 2.2 and Figure 2.29). When a message is received by a branch node, it is forwarded in the following way depending on the bit value:

1) If the first bit is set to 1, the message is forwarded to all nodes (*broadcast*).

2) If the second bit is set to 1, the message is forwarded to the downstream branch node.

3) If the third bit is set to 1, the message is forwarded to the upstream branch node.

4) If the fourth bit is set to 1, the message is forwarded to the elected leaf nodes.

Message type	1 2 3 4	5 6 7 8
Transmission to all branch nodes (*broadcast* mode)	1 – – –	In compliance with OLSR
Transmission to @ChainUP	– 1 – –	
Transmission to @ChainDO	– – 1 –	
Transmission to the elected leaf nodes	– – – 1	

Table 2.2. *Modification for CBL of the "message type" defined in the OLSR protocol in order to take into account the different possible retransmission cases*

The direction information in "Dir" messages makes it possible for the virtual infrastructure structure obtained by CBL to provide a solution to the four communication scenarios defined in ETSI (2013) for automotive applications specified by the *European Telecommunications Standards Institute* (ETSI): point-to-point, point-to-multipoint, *GeoAnyCast* and *GeoBroadcast* (section 2.5.1.2, Figure 2.21). For a CBL-OLSR implementation, point-to-point and point-to-multipoint communication scenarios are performed by the *unicast* and *multicast* mechanisms of the OLSR protocol, respectively. The *GeoAnyCast* and *GeoBroadcast* scenarios could be implemented jointly with the position information contained in the routing table of neighboring nodes, the direction information of "Dir" messages, with the encoding of a new type of communication mode in the "Message type" field using the remaining free bits of OLSR (Table 2.2) and resorting to the CBL structure. Consider an area of geographic interest to which a source node wants to send a message (yellow circle in Figure 2.21). The source node would identify the relative position upstream or downstream of this area of interest. It would send the application message (containing a new field identifying the position and size of the geographic area) to the node of its chain with the "Dir" direction information set up so that the message would be forwarded in the correct direction. Each time the message is received by a relay branch node, when the "Message type" field indicates a communication according to one of the *GeoAnyCast* and *GeoBroadcast* scenarios, the node would verify if the position of the next branch node lies within the targeted area. For a *GeoAnyCast* scenario, when this area is reached, the relay node would overwrite the *Time To Live* (TTL) field of the message with a value of 1 validating a final retransmission. For a *GeoBroadcast* scenario, when this area is reached, the relay node would overwrite the “Dir” field of the message with the value “1001XXXX” enabling local broadcast to all branch and leaf nodes. Each branch node receiving this message would rebroadcast it in case it is located in the target area.

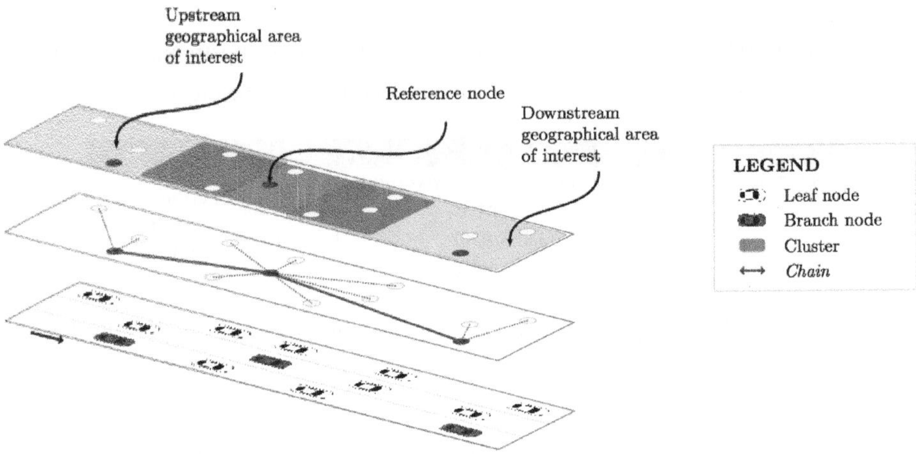

Figure 2.29. *CBL Nodes and functional links: at the bottom, an illustration of road traffic; in the middle, CBL nodes corresponding to vehicles and links between nodes; on top, areas of interest upstream and downstream of a particular branch node. For a color version of this figure, see www.iste.co.uk/wahl/design.zip*

The implementation of CBL in the OLSR protocol also required several modifications to the HELLO messages to add information about position, node type (branch or leaf) and speed, direction angle as well as the choice of nodes regarding the chain structure (@BranchChoice, @ChainUP and @ChainDO).

2.5.3.3. *Path search method*

The path search principle of the OLSR protocol was not changed by the CBL-OLSR implementation. As in the OLSR protocol, TC messages are sent by relay nodes. Each relay node writes in its TC messages a list of neighboring nodes that have selected it. Therefore, the branch nodes of CBL-OLSR generate TC messages containing a list of their attached leaf nodes. These messages are relayed throughout the network in *broadcast* mode with the chain.

2.5.3.4. *Changes to neighbor tables*

This section concludes the algorithm overview with a description of the changes that were made to the header and body of HELLO messages when implementing

CBL in OLSR. The routing tables for each node were modified to add additional information about neighboring nodes. This table contains 13 fields, the encoding of which is shown in Table 2.3.

Fields	1	2	3	4	5	6	7	8	9	10	11	12	13
Number of bits used for the encoding	Address	Link	Date 1	Date 2	Position X	Position Y	Speed	Angle	Type	Elected	N.Br.Choice	N.ChainDO	N.ChainUP
	32	2	32	32	16	16	8	8	1	1	32	32	32

Table 2.3. *Encoding the one-hop neighbor table of a node*

Link code	1	2	3	4	5	6	7	8
@ChainUP in 1st position	–	–	–	1	Part of the encoding compliant with that of OLSR			
@ChainUP not elected	–	–	–	0				
@ChainDO in 1st position	–	–	1	0				
@ChainDO in 2nd position	–	–	1	1				
@ChainDO not elected	–	–	0	–				
@BranchChoice in 1st position	–	1	0	0				
@BranchChoice in 2nd position	–	1	1	0				
@BranchChoice in 2nd position	–	1	0	1				
@BranchChoice in 3rd position	–	1	1	1				
@BranchChoice not elected	–	0	–	–				

Table 2.4. *Proposed encoding for the Link type, for use of the CBL schema, which includes the order of appearance of the choice of relay nodes @BranchChoice, @ChainUP, and @ChainDO*

2.6. Conclusion

In this chapter, we have illustrated the application of a functional approach to the design of a routing protocol for VANETs. The starting point is a set of functional requirements associated with routing datagrams between vehicles in the road environment. In this regard, we first put forward the need to bring out, through an appropriate selection of vehicles assuming a particular function, a structure similar to that which would have been obtained if access points of an infrastructure had been deployed along the road. Next, we considered the need to form disjointed groups between vehicles in order to facilitate the deployment of cooperative applications that require it. Finally, we listed the needs related to the implementation of all kinds of communications, including those oriented to certain areas of road traffic, including those defined by the *GeoNetworking* standard.

The proposed idea is to start from an analysis of the road configuration and the dynamics of urban traffic in order to set up the most suitable organization between vehicles in order to best meet the identified requirements. With the *Chain-Branch-Leaf* (CBL) *clustering* method, which creates a virtual infrastructure in each direction of traffic on a road, we were able to illustrate our approach by virtue of a case study. This infrastructure connects all vehicles through a backbone (chain) formed by elected nodes, the branch nodes, which operate as access points. All nonelected nodes, leaf nodes, when they move in a portion of the road covered by a branch node, choose the latter as a relay for their traffic. The CBL chain structure allows for the introduction of the targeted communication services. First of all, it is possible to define the directions in which messages are broadcast (within the group, upstream of the traffic, downstream, and to the whole VANET) in order to confine the impact of this broadcast. Secondly, it makes it possible to implement the four communication modes specified by the *GeoNetworking* standard (point-to-point, point-to-multipoint, *GeoAnyCast*, *GeoBroadcast*). Finally, it creates disjointed groups around the branch nodes adapted to the operation of cooperative applications. A description of the integration of CBL into the OLSR protocol showed that this structuring method can be perfectly integrated into existing routing protocols, provided that the latter make it possible to periodically transmit a message for discovering the direct neighborhood of each node. With respect to OLSR, the modification to the HELLO messages as well as the appropriate treatments in order to utilize CBL from OLSR have been described. Nevertheless, this case study presented with OLSR should not affect the perspective that CBL can be used with other routing protocols.

We have observed that this functional design approach, which relies heavily on geometric intuition in the distributed implementation of the virtual infrastructure, does not necessarily lead to an optimal result. Therefore, based on the simulation of a network of vehicles operating the protocol in several road traffic configurations, we were able to observe malfunctions that were solved thus resulting in the gradual improvement of the routing protocol. We mentioned in particular that the simulation had allowed us to point out reasons for the creation of parallel chains and the means to avoid them. We were also able to observe that certain parameters could contribute to a greater stability of the backbone (longer branch status duration) or, on the contrary, a greater stability of the disjoint clusters that were formed (longer average duration of leaf-branch connection). Consequently, variants of the protocol were devised that could be alternately implemented depending on the contexts in which they would make a greater contribution to performance.

Following this chapter on functional design, we will describe, the simulation models and tools that were implemented during the design and evaluation process of the routing protocol of the case study in Chapter 3. A detailed description of the scenarios developed will be provided in order to facilitate the reproducibility of our experiments.

2.7. References

AlMheiri, S.M. and AlQamzi, H.S. (2015). MANETs and VANETs clustering algorithms: A survey. In *2015 IEEE 8th GCC Conference & Exhibition*, Muscat. doi: 10.1109/IEEEGCC.2015.7060048.

Amjad, Z. and Song, W.-C. (2015). Road aware QoS routing in VANETs. In *2015 17th Asia-Pacific Network Operations and Management Symposium (APNOMS)*. IEEE, Busan. doi: 10.1109/AP-NOMS.2015.7275415.

Arkian, H.R., Atani, R.E., Kamali, S. (2014). FcVcA: A fuzzy clustering-based vehicular cloud architecture. In *2014 7th International Workshop on Communication Technologies for Vehicles (Nets4Cars-Fall)*. IEEE, St. Petersburg. doi: 10.1109/Nets4CarsFall. 2014.7000907.

Arkian, H.R., Atani, R.E., Pourkhalili, A., Kamali, S. (2015). A stable clustering scheme based on adaptive multiple metric in vehicular ad-hoc networks. *Journal of Information Science & Engineering*, 31(2), 361–386.

Avcil, M.N. and Soyturk, M. (2015). ReSCUE: Relatively stable clustering for unbiased environments in VANETs. In *2015 International Wireless Communications and Mobile Computing Conference (IWCMC)*. IEEE, Dubrovnik. doi: 10.1109/iwcmc.2015.7289227.

Ayaida, M., Fouchal, H., Afilal, L., Ghamri-Doudane, Y. (2012). A comparison of reactive, grid and hierarchical location-based services for VANETs. In *2012 IEEE Vehicular Technology Conference (VTC Fall)*, Quebec City. doi: 10.1109/VTCFall.2012.6398920.

Azizian, M., Cherkaoui, S., Hafid, A.S. (2016). DCEV: A distributed cluster formation for VANET based on end-to-end realtive mobility. In *2016 International Wireless Communications and Mobile Computing Conference (IWCMC)*. IEEE, Paphos. doi: 10.1109/iwcmc.2016.7577072.

Badis, H. and Al Agha, K. (2004). QOLSR multi-path routing for mobile ad hoc networks based on multiple metrics: Bandwidth and delay. In *2004 IEEE 59th Vehicular Technology Conference, VTC 2004-Spring (IEEE Cat. No. 04CH37514)*, Milan. doi: 10.1109/vetecs.2004.1390660.

Badis, H. and Al Agha, K. (2005). QOLSR, QoS routing for ad hoc wireless networks using OLSR. *Transactions on Emerging Telecommunications Technologies*, 16(5), 427–442. doi: 10.1002/ett.1067.

Bali, R.S., Kumar, N., Rodrigues, J.J. (2014). Clustering in vehicular ad hoc networks: Taxonomy, challenges and solutions. *Vehicular Communications*, 1(3), 134–152. doi: 10.1016/j.vehcom.2014.05.004.

Basagni, S., Chlamtac, I., Syrotiuk, V.R., Woodward, B.A. (1998). A distance routing effect algorithm for mobility (DREAM). In *Proceedings of the 4th Annual ACM/IEEE International Conference on Mobile Computing and Networking*. ACM, Dallas. doi: 10.1145/ 288235.288254.

Beijar, N. (2002). Zone routing protocol (ZRP). *Networking Laboratory, Helsinki University of Technology*, 9, 1–12 [Online]. Available at: https://www.netlab.tkk.fi/opetus/s38030/k02/Papers/ 08-Nicklas.pdf.

Belamri, F., Boulfekhar, S., Aissani, D. (2021). A survey on QoS routing protocols in Vehicular Ad hoc Network (VANET). *Telecommun. Syst.*, 78, 117–153. doi: 10.1007/s11235-021-00797-8.

Beylot, A.-L. and Labiod, H. (2013). *Vehicular Networks: Models and Algorithms*. ISTE Ltd, London, and John Wiley & Sons, New York.

Bitam, S. and Mellouk, A. (2011). QoS swarm bee routing protocol for vehicular ad hoc networks. In *2011 IEEE International Conference on Communications (ICC)*, Kyoto. doi: 10.1109/ icc.2011.5963424.

Blum, J., Eskandarian, A., Hoffman, L. (2004). Challenges of intervehicle ad hoc networks. *IEEE Transactions on Intelligent Transportation Systems*, 5(4), 347–351. doi: 10.1109/TITS.2004.838218.

Boussoufa-Lahlah, S., Semchedine, F., Bouallouche-Medjkoune, L. (2018). Geographic routing protocols for Vehicular Ad hoc NETworks (VANETs): A survey. *Vehicular Communications*, 11, 20–31. doi: 10.1016/j.vehcom.2018.01.006.

Cha, S.-H., Lee, J.-E., Ryu, M. (2016). Directed broadcasting with mobility prediction for vehicular sensor networks. *International Journal of Distributed Sensor Networks*, 12(7), 1–9. doi: 10.1177/1550147716657930.

Chakchouk, N. (2015). A survey on opportunistic routing in wireless communication networks. *IEEE Communications Surveys Tutorials*, 17(4), 2214–2241.

Chen, T.-W. and Gerla, M. (1998). Global state routing: A new routing scheme for ad-hoc wireless networks. In *ICC'98. 1998 IEEE International Conference on Communications. Conference Record, SUPERCOMM'98, (Cat. No. 98CH36220)*, Atlanta. doi: 10.1109/icc.1998.682615.

Clausen, T., Jacquet, P., Adjih, C., Laouiti, A., Minet, P., Muhlethaler, P., Qayyum, A., Viennot, L. (2003). Optimized link state routing protocol (OLSR), RFC 3626. ACM, Brussels. doi: 10.17487/RFC3626.

Cooper, C., Franklin, D., Ros, M., Safaei, F., Abolhasan, M. (2017). A comparative survey of VANET clustering techniques. *IEEE Communications Surveys & Tutorials*, 19(1), 657–681. doi: 10.1109/COMST.2016.2611524.

Daeinabi, A., Rahbar, A.G.P., Khademzadeh, A. (2011). VWCA: An efficient clustering algorithm in vehicular ad hoc networks. *Journal of Network and Computer Applications*, 34(1), 207–222. doi: 10.1016/j.jnca.2010.07.016.

Dagdeviren, O. and Erciyes, K. (eds) (2006). A distributed backbone formation algorithm for mobile ad hoc networks. In *International Symposium on Parallel and Distributed Processing and Applications*. Springer, Berlin. doi: 10.1007/11946441_24.

Daknou, E., Thaalbi, M., Tabbane, N. (2015). A fast clustering algorithm for VANETs. In *Proceedings of the 13th International Conference on Advances in Mobile Computing and Multimedia*. ACM, Brussels. doi: 10.1145/2837126.2837147.

Delot, T., Cenerario, N., Ilarri, S. (2010). Vehicular event sharing with a mobile peer-to-peer architecture. *Transportation Research Part C: Emerging Technologies*, 18(4), 584–598. doi: 10.1016/j.trc.2009.12.003.

Di Felice, M., Bedogni, L., Bononi, L. (2011). Dynamic backbone for fast information delivery invehicular ad hoc networks: An evaluation study. In *PE-WASUN'11: Proceedings of the 8th ACM Symposium on Performance Evaluation of Wireless Ad Hoc, Sensor, and Ubiquitous Networks*. ACM, Miami. doi: 10.1145/2069063.2069065.

ETSI (2011a). Intelligent transport systems (ITS); Vehicular communications; GeoNetworking; Part 5: Transport protocols; Sub-part 1: Basic transport protocol. Technical Specification ETSI TS 102 636-5-1 V1.1.1 (2011-02), ETSI Technical Committee Intelligent Transport System, Sophia Antipolis [Online]. Available at: https://www.etsi.org/standards.

ETSI (2011b). Intelligent transport systems (ITS); Vehicular communications; GeoNetworking; Part 6: Internet integration; Sub-part 1: Transmission of IPv6 packets over GeoNetworking protocols. Technical Specification ETSI TS 102 636-6-1 V1.1.1 (2011-03), ETSI Technical Committee Intelligent Transport System, Sophia Antipolis [Online]. Available at: https://www.etsi.org/standards.

ETSI (2013). Intelligent transport systems (ITS); Vehicular communications; GeoNetworking; Part 2: Scenarios. European Standard ETSI EN 302 636-2 V1.2.0 (2013-07), ETSI Technical Committee Intelligent Transport System, Sophia Antipolis [Online]. Available at: https://www.etsi.org/standards.

ETSI (2019a). Intelligent transport systems (ITS); Vehicular communications; Basic set of applications; Part 2: Specification of cooperative awareness basic service. European Standard ETSI EN 302 367-2 V1.4.1 (2019-04), ETSI Technical Committee Intelligent Transport System, Sophia Antipolis [Online]. Available at: https://www.etsi.org/standards.

ETSI (2019b). Intelligent transport systems (ITS); Vehicular communications; Basic set of applications; Part 3: Specifications of decentralized environmental notification basic service. European Standard ETSI EN 302 367-3 V1.3.1 (2019-04), ETSI Technical Committee Intelligent Transport System, Sophia Antipolis [Online]. Available at: https://www.etsi.org/standards.

Fan, W., Shi, Y., Chen, S., Zou, L. (2011). A mobility metrics based dynamic clustering algorithm for VANETs. In *ICCTA 2011*. IET, Beijing. doi: 10.1049/cp.2011.0769.

Frikha, M. (2010). *Ad Hoc Networks: Routing, QoS and Optimization*. ISTE Ltd, London, and John Wiley & Sons, New York.

Gantsou, D., Sondi, P., Hanafi, S. (2009). Revisiting multipoint relay selection in the optimized link state routing protocol. *International Journal of Communication Networks and Distributed Systems*, 2(1), 4–15. doi: 10.1504/ IJCNDS.2009.021691.

Ge, Y., Kunz, T., Lamont, L. (2003). Quality of service routing in ad-hoc networks using OLSR. In *Proceedings of the 36th Annual Hawaii International Conference on System Sciences (HICSS'03)*. IEEE, Big Island, Hawaii. doi: 10.1109/hicss.2003.1174847.

Guangyu, P., Gerla, M., Tsu-Wei, C. (2000). Fisheye state routing: A routing scheme for ad hoc wireless networks. In *2000 IEEE International Conference on Communications. ICC 2000. Global Convergence Through Communications*, 1, 70–74, New Orleans.

Hadded, M., Muhlethaler, P., Laouiti, A., Saidane, L.A. (eds) (2017). A novel angle-based clustering algorithm for vehicular ad hoc networks. In *Vehicular Ad-Hoc Networks for Smart Cities*. Springer, Berlin. doi: 10.1007/ 978-981-10-3503-6_3.

Harrabi, S., Jaffar, I.B., Ghedira, K. (2016). Novel optimized routing scheme for VANETs. *Procedia Computer Science*, 98, 32–39. doi: 10.1016/j.procs.2016.09.008.

Hnatyshin, V. and Asenov, H. (2010). Design and implementation of an OPNET model for simulating GeoAODV MANET routing protocol. In *Proc. of the OPNETWORK 2010 International Conference, Session: Wireless Ad Hoc and Wireless Personal Area Networks*, Washington.

Houssaini, Z.S., Zaimi, I., Oumsis, M., Ouatik, S.E.A. (2017). GPSR+Predict: An enhancement for GPSR to make smart routing decision by anticipating movement of vehicles in VANETs. *Adv. Sci. Technol. Eng. Syst. J.*, 2(3), 137–146. doi: 10.25046/aj020318.

Hussey, R., Huff, E., Shinwari, Z., Hnatyshin, V. (2013). A comparative study of proactive and reactive geographical routing protocols for MANET. In *Proc. of the International Conference on Wireless Networks (ICWN)*. IAENG, London.

Jemaa, I.B. (2014). Multicast communications for cooperative vehicular systems. PhD Thesis, Mines ParisTech, Paris [Online]. Available at: https://hal.inria.fr/tel-01101679.

Jerbi, M., Meraihi, R., Senouci, S.-M., Ghamri-Doudane, Y. (2006). GyTAR: Improved greedy traffic aware routing protocol for vehicular ad hoc networks in city environments. In *VANET'06: Proceedings of the 3rd International Workshop on Vehicular Ad Hoc Networks*. ACM, Los Angeles. doi: 10.1145/1161064.1161080.

Johnson, D., Hu, Y., Maltz, D. (2007). The Dynamic Source Routing protocol (DSR) for mobile ad hoc networks for IPv4 (RFC 4728). Report, IETF [Online]. doi: 10.17487/RFC4728.

Kalivaradhan, R.B. and Thangavelu, A. (2010). A simulated modeling approach towards providing adaptive QoS for vehicular safety services over VANET. *International Journal of Research and Reviews in Computer Science*, 1(4), 110–116.

Karp, B. and Kung, H.-T. (2000). GPSR: Greedy perimeter stateless routing for wireless networks. In *MobiCom'00: Proceedings of the 6th Annual International Conference on Mobile Computing and Networking*. ACM, Boston. doi: 10.1145/345910.345953.

Katiyar, A., Singh, D., Yadav, R. (2020). State-of-the-art approach to clustering protocols in VANET: A survey. *Wireless Networks*, 26, 5307–5336. doi: 10.1007/ s11276-020-02392-2.

Ko, Y.-B. and Vaidya, N.H. (2000). Location-aided routing (LAR) in mobile ad hoc networks. *Wireless Networks*, 6(4), 307–321. doi: 10.1023/A:1019106118419.

Kuklinski, S. and Wolny, G. (2009). Density based clustering algorithm for VANETs. In *2009 5th International Conference on Testbeds and Research Infrastructures for the Development of Networks & Communities and Workshops*. IEEE, Washington. doi: 10.1109/tridentcom.2009.4976256.

Lebedev, D. (2006). Ad hoc networks: Study of protocol behaviour. PhD Thesis, Laboratoire d'informatique de l'École polytechnique (LIX), Paris [Online]. Available at: https://www.lix.polytechnique.fr/Labo/Dmitri.Lebedev/lebedev_these.pdf.

Lee, K.C., Le, M., Harri, J., Gerla, M. (2008). LOUVRE: Landmark overlays for urban vehicular routing environments. In *2008 IEEE 68th Vehicular Technology Conference*, Calgary. doi: 10.1109/vetecf.2008.447.

Leguay, J., Conan, V., Friedman, T. (2006). QoS routing in OLSR with several classes of service. In *Fourth Annual IEEE International Conference on Pervasive Computing and Communications Workshops (PERCOMW'06)*, Pisa. doi: 0.1109/percomw.2006.115.

Li, G. (2015). Adaptive and opportunistic QoS-based routing protocol in VANETs. PhD Thesis, Université Paris Sud, Paris [Online]. Available at: https://tel.archives-ouvertes.fr/tel-01244591.

Li, Y., Jin, D., Wang, Z., Zeng, L., Chen, S. (2013). Exponential and power law distribution of contact duration in urban vehicular ad hoc networks. *IEEE Signal Processing Letters*, 20(1), 110–113. doi: 10.1109/lsp.2012.2231412.

Li, G., Boukhatem, L., Martin, S. (2015). An intersection-based QoS routing in vehicular ad hoc networks. *Mobile Networks and Applications*, 20(2), 268–284. doi: 10.1007/s11036-015-0577-4.

Lim, H. and Kim, C. (2001). Flooding in wireless ad hoc networks. *Computer Communications*, 24(3), 353–363. doi: 10.1016/ S0140-3664(00)00233-4.

Lin, Y.-W. and Huang, G.-T. (2012). Optimal next hop selection for VANET routing. In *7th International Conference on Communications and Networking in China*. Kunming. doi: 10.1109/china-com.2012.6417556.

Lin, Q., Li, C., Wang, X., Zhu, L. (2013). A three-dimensional scenario oriented routing protocol in vehicular ad hoc networks. In *2013 IEEE 77th Vehicular Technology Conference (VTC Spring)*, IEEE, Dresden. doi: 10.1109/vtcspring.2013.6691848.

Liu, C. and Chigan, C. (2012). RPB-MD: Providing robust message dissemination for vehicular ad hoc networks. *Ad Hoc Networks*, 10(3), 497–511. doi: 10.1016/j.adhoc.2011.09.003.

Luo, Y., Zhang, W., Hu, Y. (2010). A new cluster based routing protocol for VANET. In *2010 Second International Conference on Networks Security, Wireless Communications and Trusted Computing, volume 1*. IEEE, Wuhan. doi: 10.1109/NSWCTC.2010.48.

Maslekar, N., Boussedjra, M., Mouzna, J., Labiod, H. (2011). C-DRIVE: Clustering based on direction in vehicular environment. In *2011 4th IFIP International Conference on New Technologies, Mobility and Security*. IEEE, Paris. doi: 10.1109/NTMS.2011.5720653.

Mehra, R., Bali, R.S., Kaur, P. (2016). Efficient clustering based OLSR routing protocol for VANET. In *2016 Symposium on Colossal Data Analysis and Networking (CDAN)*. IEEE, Indore. doi: 10.1109/CDAN.2016.7570915.

Morales, M.M.C., Hong, C.S., Bang, Y.-C. (2011). An adaptable mobility-aware clustering algorithm in vehicular networks. In *2011 13th Asia-Pacific Network Operations and Management Symposium*. IEEE, Taipei. doi: 10.1109/apnoms.2011.6077004.

Mukunthan, A. (2015). Routing and propagation in urban vehicular ad hoc networks, PhD Thesis, University of Wollongong, Wollongong [Online]. Available at: http://ro.uow.edu.au/ theses/4558/.

Naumov, V. and Gross, T.R. (2007). Connectivity-aware routing (CAR) in vehicular ad-hoc networks. In *INFOCOM 2007. 26th IEEE International Conference on Computer Communications*, Anchorage. doi: 10.1109/INFCOM.2007.223.

Naumov, V., Baumann, R., Gross, T. (2006). An evaluation of inter-vehicle ad hoc networks based on realistic vehicular traces. In *MobiHoc'06: Proceedings of the 7th ACM International Symposium on Mobile Ad Hoc Networking and Computing*, Florence. doi: 10.1145/1132905.1132918.

Navidi, W. and Camp, T. (2004). Stationary distributions for the random waypoint mobility model. *IEEE Transactions on Mobile Computing*, 3(1), 99–108. doi: 10.1109/TMC.2004.1261820.

Nguyen, D.-Q. and Minet, P. (2006). QoS support and OLSR routing in a mobile ad hoc network. In *International Conference on Networking, International Conference on Systems and International Conference on Mobile Communications and Learning Technologies (ICN/ICONS/MCL'06)*. IEEE, Morne. doi: 10.1109/icniconsmcl.2006.178.

Ogier, R., Templin, F., Lewis, M. (2004). Topology dissemination based on reverse-path forwarding (TBRPF). *RFC 3684*. doi: 10.17487/RFC3684.

Olariu, S. and Weigle, M.C. (2009). *Vehicular Networks: From Theory to Practice*. Chapman & Hall/CRC, London.

Perkins, C.E. and Bhagwat, P. (1994). Highly dynamic destination-sequenced distance-vector routing (DSDV) for mobile computers. In *ACM SIGCOMM Computer Communication Review, volume 24*. ACM, London. doi: 10.1145/190314.190336.

Perkins, C.E., Belding-Royer, E., Das, S. (2003). Ad-hoc On-Demand Distance Vector (AODV) Routing, *RFC 3561*. doi: 10.17487/RFC3561.

Qayyum, A., Viennot, L., Laouiti, A. (2000). Multipoint relaying: An efficient technique for flooding in mobile wireless networks. Research Report, RR-3898, INRIA, Paris [Online]. Available at: https://hal.inria.fr/inria-00072756.

Rahbar, H., Naik, K., Nayak, A. (2010). DTSG: Dynamic time-stable geocast routing in vehicular ad hoc networks. In *2010 The 9th IFIP Annual Mediterranean Ad Hoc Networking Workshop (Med-Hoc-Net)*. IEEE, Juan Les Pins. doi: 10.1109/medhocnet.2010.5546872, organization=IEEE.

Rawashdeh, Z.Y. and Mahmud, S.M. (2012). A novel algorithm to form stable clusters in vehicular ad hoc networks on highways. *EURASIP Journal on Wireless Communications and Networking*, 1, 15. doi: 10.1186/1687-1499-2012-15.

Ren, M., Khoukhi, L., Labiod, H., Zhang, J., Vèque, V. (2017). A mobility-based scheme for dynamic clustering in vehicular ad-hoc networks (VANETs). *Vehicular Communications*, 9, 233–241. doi: 10.1016/j.vehcom.2016.12.003.

Rice, J.A. (2006). *Mathematical Statistics and Data Analysis*. Duxbury Press, Belmont.

Rivoirard, L. (2018). Modèle d'auto-organisation pour les protocoles de routage dans les réseaux ad hoc de véhicules : application à la perception élargie et à la localisation coopératives. PhD Thesis, Université de Lille, Lille [Online]. Available at: http://www.theses.fr/2018LIL1I041.

Rivoirard, L., Wahl, M., Sondi, P., Berbineau, M., Gruyer, D. (2016). Using real-world car traffic dataset in vehicular ad hoc network performance evaluation. *International Journal of Advanced Computer Science and Applications (IJACSA)*, 7(12), 390–398. doi: 10.14569/IJACSA.2016.071251.

Rivoirard, L., Wahl, M., Sondi, P., Berbineau, M., Gruyer, D. (2018). Chain-Branch-Leaf: A clustering scheme for vehicular networks using only V2V communications. *Ad Hoc Networks*, 68, 70–84. doi: 10.1016/j.adhoc.2017.10.007.

Rivoirard, L., Wahl, M., Sondi, P. (2020). Multipoint relaying versus chain-branch-leaf clustering performance in optimized link state routing-based vehicular ad hoc networks. *IEEE Transactions on Intelligent Transportation Systems*, 21(3), 1034–1043. doi: 10.1109/TITS.2019.2900767.

Rohani, M., Gingras, D., Gruyer, D. (2016). A novel approach for improved vehicular positioning using cooperative map matching and dynamic base station DGPS concept. *IEEE Transactions on Intelligent Transportation Systems*, 17(1), 230–239. doi: 10.1109/tits.2015.2465141.

Sahu, P.K., Wu, E.H.-K., Sahoo, J., Gerla, M. (2013). BAHG: Back-bone-assisted hop greedy routing for VANET's city environments. *IEEE Transactions on Intelligent Transportation Systems*, 14(1), 199–213. doi: 10.1109/tits.2012.2212189.

Sarr, C. (2007). De l'apport d'une évaluation précise des ressources pour la Qualité de Service des réseaux ad hoc basés sur IEEE 802. 11. PhD Thesis, INSA, Villeurbanne [Online]. Available at: http://www.theses.fr/2007ISAL0046.

Seet, B.-C., Liu, G., Lee, B.-S., Foh, C.-H., Wong, K.-J., Lee, K.-K. (2004), A-STAR: A mobile ad hoc routing strategy for metropolis vehicular communications. In *Networking 2004*. Springer, Berlin. doi: 10.1007/978-3-540-24693-0_81.

Senouci, O., Harous, S., Aliouat, Z. (2020). Survey on vehicular ad hoc networks clustering algorithms: Overview, taxonomy, challenges, and open research issues. *International Journal of Communication Systems*, 33(11), e4402. doi: 10.1002/dac.4402.

Sethi, V. and Chand, N. (2017). A destination based routing protocol for context based clusters in VANET. *Communications and Network*, 9(03), 179. doi: 10.4236/cn.2017.93013.

Shi, T., Wang, N., Xiong, P. (2009). An on-demand virtual backbone broadcast protocol for VANET. In *2009 1st International Conference on Information Science and Engineering (ICISE)*. IEEE, Nanjing. doi: 10.1109/ICISE.2009.287.

Sondi, P. (2010). Le routage à qualité de service dans les réseaux mobiles ad hoc (Routing with quality of service support in mobile ad hoc networks). PhD Thesis, Université Polytechnique Hauts-de-France, Valenciennes [Online]. Available at: https://tel.archives-ouvertes.fr/tel-00551243.

Sondi, P., Gantsou, D., Lecomte, S. (2010). Performance evaluation of multimedia applications over an OLSR-based mobile ad hoc network using OPNET. In *12th International Conference on Computer Modelling and Simulation (UKSim)*. IEEE, Cambridge. doi: 10.1109/uk-sim. 2010.109.

Sondi, P., Gantsou, D., Lecomte, S. (2013). Design guidelines for quality of service support in optimized link state routing-based mobile ad hoc networks. *Ad Hoc Networks*, 11(1), 298–323. doi: 10.1016/j.adhoc.2012.06.001.

Song, T., Xia, W., Song, T., Shen, L. (2010). A cluster-based directional routing protocol in VANET. In *2010 IEEE 12th International Conference on Communication Technology*, Nanjing. doi: 10.1109/icct.2010.5689132.

Souza, E., Nikolaidis, I., Gburzynski, P. (2010). A new aggregate local mobility (ALM) clustering algorithm for VANETs. In *2010 IEEE International Conference on Communications*, Cape Town. doi: 10.1109/icc.2010.5501789.

Sun, Y., Luo, S., Dai, Q., Ji, Y. (2015). An adaptive routing protocol based on QoS and vehicular density in urban VANETs. *International Journal of Distributed Sensor Networks*, 11(6), 631092. doi: 10.1155/2015/631092.

Tian, D., Wang, Y., Lu, G., Yu, G. (2010). A VANETs routing algorithm based on Euclidean distance clustering. In *2010 2nd International Conference on Future Computer and Communication*. IEEE, Wuhan. doi: 10.1109/ICFCC.2010.5497808.

Togou, M.A., Hafid, A., Sahu, P.K. (2014). A stable minimum velocity CDS-based virtual backbone for VANET in city environment. In *39th Annual IEEE Conference on Local Computer Networks*, Edmonton. doi: 10.1109/LCN.014.6925829.

Ucar, S., Ergen, S.C., Ozkasap, O. (2013). VMaSC: Vehicular multi-hop algorithm for stable clustering in vehicular ad hoc networks. In *2013 IEEE Wireless Communications and Networking Conference (WCNC)*, Shanghai. doi: 10.1109/ wcnc.2013.6554933.

Vèque, V. and Johnen, C. (2012). Hiérarchisation dans les réseaux ad hoc de véhicules. *8e journées francophones mobilité et ubiquité, UBIMOB 2012*. Éditions Cépaduès [Online]. Available at: https://hal.archives-ouvertes.fr/ hal-00781267.

Vèque, V., Kaisser, F., Johnen, C., Busson, A. (2013). CONVOY: A new cluster-based routing protocol for vehicular networks. In *Vehicular Networks: Models and Algorithms*, Beylot, A.-L. and Labiod, H. (eds). ISTE Ltd, London, and John Wiley & Sons, New York.

Wahab, O.A., Otrok, H., Mourad, A. (2013). VANET QoS-OLSR: QoS-based clustering protocol for vehicular ad hoc networks. *Computer Communications*, 36(13), 1422–1435. doi: 10.1016/j.comcom.2013.07.003.

Wahl, M., Sondi, P., Rivoirard, L. (2021). Enhanced CBL clustering performance versus GRP, OLSR and AODV in vehicular ad hoc networks. *Telecommunication Systems*, 76, 525–540. doi: 10.1007/s11235-020-00734-1.

Wang, Z. and Crowcroft, J. (1996). Quality-of-service routing for supporting multimedia applications. *IEEE Journal on Selected Areas in Communications*, 14(7), 1228–1234.

Wang, Z., Liu, L., Zhou, M., Ansari, N. (2008). A position-based clustering technique for ad hoc intervehicle communication. *IEEE Transactions on Systems, Man, and Cybernetics, Part C (Applications and Reviews)*, 38(2), 201–208. doi: 10.1109/TSMCC.2007.913917.

Wu, C., Ohzahata, S., Kato, T. (2013a). Can we generate efficient routes by using only beacons? Backbone routing in VANETs. In *2013 IEEE 24th Annual International Symposium on Personal, Indoor, and Mobile Radio Communications (PIMRC)*, London. doi: 10.1109/pimrc.2013.6666648.

Wu, C., Ohzahata, S., Kato, T. (2013b). Data dissemination with dynamic backbone selection in vehicular ad hoc networks. In *2013 IEEE 78th Vehicular Technology Conference (VTC Fall)*, Las Vegas. doi: 10.1109/vtcfall.2013.6692152.

Wu, C., Ohzahata, S., Ji, Y., Kato, T. (2016). How to utilize interflow network coding in VANETs: A backbone-based approach. *IEEE Transactions on Intelligent Transportation Systems*, 17(8), 2223–2237. doi: 10.1109/tits.2016. 2516027.

Xiang, Y., Liu, Z., Liu, R., Sun, W., Wang, W. (2013). GeoSVR: A map-based stateless VANET routing. *Ad Hoc Networks*, 11(7), 2125–2135. doi: 10.1109/anthology.2013.6784953.

Xiaonan, W. and Huanyan, Q. (2014). Constructing a VANET based on cluster chains. *International Journal of Communication Systems*, 27(11), 2497–2517. doi: 10.1002/dac.2484.

Yan, G., Rawat, D.B., Bista, B.B. (2012). Provisioning vehicular ad hoc networks with quality of service. *International Journal of Space-Based and Situated Computing*, 2(2), 104–111. doi: 10.1504/ijssc.2012.047467.

Yélémou, T. (2012). Utilisation du taux d'erreur binaire pour améliorer la qualité de service dans les réseaux ad hoc. PhD Thesis, Université de Poitiers, Poitiers [Online]. Available at: www.theses.fr/2012POIT2317.

Zhao, J. and Cao, G. (2008). VADD: Vehicle-assisted data delivery in vehicular ad hoc networks. *IEEE Transactions on Vehicular Technology*, 57(3), 1910–1922. doi: 10.1109/infocom.2006.298.

Zheng, J., Wu, Y., Xu, Z., Lin, X. (2014). A reliable routing protocol based on QoS for VANET. In *The 7th IEEE/International Conference on Advanced Infocomm Technology*, Fuzhou. doi: 10.1109/icait.2014.7019524.

Zhiyuan, L. (2009). Geographic routing protocol and simulation. In *Second International Workshop on Computer Science and Engineering (WCSE), volume 2*. IEEE, Qingdao. doi: 10.1109/WCSE.2009.840.

3

Models and Simulations for Performance Analysis

3.1. Introduction

To determine whether a protocol will be useful once implemented, it should be evaluated under similar conditions or close to those of a ground truth, without having to deal with the vicissitudes of a real situation, whose causes and consequences on the results obtained would be difficult to discriminate or even be uninterpretable. However, what is this ground truth that will impact the modeling and simulation conditions?

For example, consider the example of the *Chain-Branch-Leaf* (CBL) routing scheme for vehicular ad hoc networks (VANETs), introduced in section 2.5 of Chapter 2. First, the CBL scheme was designed to be integrated into an open wireless ad hoc communication network, where the spatial extent, the number of mobile nodes and even their density (number of nodes per km^2) are unknown. This communication network is a VANET. The CBL scheme could therefore be a priori executed in different road contexts (streets, roads, expressways and highways) and in various environments presenting different physical and regional geographies (plains, mountains, hillsides, cities, villages, etc.). The presence of intersections (streets and roads) or their non-existence (highways and expressways), the length of the lanes and their bends (long and almost straight lanes for highways, shorter and sometimes in narrow bends or broken lines for streets), the presence or the lack of buildings, vegetation or any drop of elevation of the road that could hinder the propagation of radio waves, for instance, are all characteristics that will have an impact on the properties of the wave propagation channels; and, for the system, on its ability to establish communications between vehicles. Therefore, an evaluation in line with the application context of the protocol will require the choice of the first road scenarios to be implemented, whether they are urban or highway; and for each of them, a radio wave propagation model consistent with the environmental constraints related to the type of road chosen in a given geographical context. Another contextual constraint must be taken into account. Mobility on these roads varies according to the time of day (for example, peak hours during commuting) and the time of year (for example, periods during or outside school holidays). Consequently, mobility models should be available in accordance to the nature of the chosen road, on the one hand, but also to its different uses, namely, with low, medium or high density. These mobility models will therefore have to take into account both the dynamic characteristics of each vehicle on the road (their speed, acceleration, steering angle, etc.) and the travel constraints imposed by a road infrastructure. We now consider the constraints on modeling the CBL scheme within a wireless communicating node. CBL is admittedly an ad hoc routing scheme of layer 3 or network layer (according to the formalism of the ISO OSI model), whose structural evaluation is possible with only taking the mobility of the nodes into account. However, assessing CBL within a communication system requires a system to be modeled. A choice must therefore be made concerning the underlying radio technology (layers 1 and 2) supporting ad hoc communications. This technology will decide how access to the communication medium and the transmission of data are achieved, as well as the theoretical maximum communication range P in free space (which appears in the definition of equation [2.3], in section 2.5.2.4). CBL does not implement all the routing mechanisms in layer 3, nor those for addressing in this layer; the CBL scheme should therefore be completed with the missing parts and protocols. For layer 7, application traffic is necessary to evaluate the operation of the communication system as a whole. Therefore, application data transfer scenarios are also to be defined. This involves, in addition to the definition of application layer (layer 7) scenarios, the use of one or more transport protocols (layer 4), such as TCP or UDP if an IP network addressing protocol is chosen in layer 3. Finally, because the objective of the person who will evaluate the CBL is not to design a network simulator, nor the set of protocols that has to be implemented to evaluate it, a simulator will also have to be chosen such that it will provide access to a set of protocols already integrated on the basis of which the modeling will be possibly built.

While modeling for the sake of simulation may seem rather complicated, directly testing the protocol on the ground is not simpler. Admittedly, some choices to be made for the simulation will have to be implemented for a field evaluation. However, the imagined protocol may not be realistic, or the model and the simulation may reveal some difficulties that could not have initially be taken into consideration on paper. The model will also ignore certain elements of the system, for example, in the case of CBL, the choice of antennas (they will be assumed to be omnidirectional and perfect), the processing of vehicle positioning information (GPS data) and its dynamics (speed data, acceleration, steering angle, etc.). On the other hand, for a real implementation with ground testing, a real-time data acquisition and storage system must be implemented in order to be able to perform the a posteriori evaluations; a sufficient number of vehicles is required for testing the behavior of the CBL scheme; a road network or an experimental track of sufficient size for the problem to be evaluated must be available. After having solved a set of experimental, economic and legal challenges posed by in situ experiments, developers can finally expect to recover data to be analyzed with the assumption, however, that all the data necessary for its evaluation have been anticipated. Vehicles, equipment and performing field tests are costly in terms of human and material resources, and difficult to implement. When it is possible to perform experiments, this is very often achieved with a limited number of vehicles, as part of European research projects on cooperative intelligent transport systems (ITSs) C-ITS (*Cooperative Intelligent Transport Systems*), such as SCOOP@F (Ollinger 2020), InterCor (InterCor 2019) and InDiD (C-Roads 2023), and finally, with the participation of car manufacturers and equipment manufacturers (OEMs) (Rivoirard 2018).

Computer modeling and simulation of an ad hoc communication system or subsystem to be evaluated is therefore, if not an alternative, a necessary prerequisite to real-situation testing, which is difficult to implement. They make it possible to evaluate a protocol in conditions similar to (or close to) those of ground truth, but not exposed to the vagaries of real situations and the dangers involved. They can be used during the development phase of the protocol to raise and solve design problems.

There is one important point that should be taken into consideration. The entire system and the constraints, and in particular environmental constraints, to which it is subject do not have to be modeled (Fishman 2001): only the essential characteristics of the system to be validated must be modeled. This requires abstracting the characteristics of the system that require performance evaluation. With respect to CBL, two evaluation approaches for the routing protocol have been devised, one structural and the other system:

– *Structural evaluation (Str)* consists of the study of the performance of the structure formed by a protocol when it creates a hierarchical topology in the network. In this approach, the layers of the Open Systems Interconnection (OSI) reference model – defined by the International Organization for Standardization (ISO) – that interact with the network layer supporting the routing protocol under consideration are idealized in order to essentially identify the impact of the latter. For the evaluation of the CBL scheme which is independent of radio technologies and applications, only modeling of node mobility and the radio range of the propagation channel is required. The expected evaluation results involve only the structural aspects of the routing protocol, such as node types (branch or leaf), the number of structures formed (chains) or the time required to form a structure.

– *System evaluation (Sys)* takes into account all the layers (according to the OSI model terminology) that are implemented by the communication network. Modeling the CBL *clustering* scheme within a communication system for VANETs then requires modeling the network layer in which the CBL scheme is integrated, that of the upper application and transport layers which support, in particular, the application profiles of users and the preparation of data packets; and finally that of the lower layers, data link and physics of radio technology. The data link layer will model accesses to the medium and a propagation model will be associated with the physical layer.

Table 3.1 summarizes the modeling requirements for a structural evaluation and a system evaluation of the CBL scheme.

Evaluation / OSI layers (1 to 7) and ground characteristics (G)	Structural	System
7 – application	–	Application model
4 – transport	–	TCP/UDP stack
3 – network	CBL scheme	IP, CBL scheme in an ad hoc network protocol model
1 and 2 – physical and data link	–	Radio communication technology interface model
G – mobile nodes	Mobility model, for a road environment to be specified	Mobility model, for a road environment to be specified
G – wave propagation model	No but maximal communication distance	Wave attenuation profile

Table 3.1. *Models required according to the type of CBL evaluation to be carried out*

In section 3.2, we specify the radio technology that will be modeled and the road context in which the system evaluations of the CBL scheme will be made in Chapter 4. Section 3.3 focuses on the families of tools for the simulation and modeling of wireless communication networks and those for the generation of mobility models implemented for the purpose of VANETs evaluation. Thirdly, starting from section 3.4, we will present the tools and models chosen for performing the structural, routing and application system evaluations of Chapter 4.

3.2. Choice of road context and radio technology

The choice of road environment will derive the choice of mobility models to be implemented and a knowledge of the types of obstacles that block the propagation of radio waves. The mobility of communication nodes and the propagation environment in which they evolve have an impact on the evaluation of ad hoc network protocols. This was in particular shown in the work of Camp et al. (2002), which compares the performance of the *Dynamic Source Routing* (DSR) reactive ad hoc routing protocol according to the stochastic mobility model used, or in the comparative study of Rhattoy and Zatni (2012) on the performance of the DSR, *Ad-hoc On Demand Distance Vector* (AODV) and *Destination-Sequenced Distance-Vector Routing* (DSDV) reactive ad hoc routing protocols according to the wave propagation model employed and the number of connections. In their summary studies on network and mobility modeling tools, Härri et al. (2009) introduce a minimum set of requirements to be taken into account (Table 3.2) for the construction of realistic mobility models, depending on whether the context of the study is urban or highway.

	Minimum requirements			Options	
Model / Context	Car-following model	Intersection	Lane change	Radio obstacles	Validation*
Highway	✓		✓		✓
Urban	✓	✓		✓	✓

* Validation of model traces achieved by computing the error with real mobility traces, or when there are none, with those of a validated model.

Table 3.2. *Minimum requirements for obtaining a realistic mobility model according to Härri et al. (2009)*

In our example, the CBL scheme has been designed to compensate for the absence of a faulty communication infrastructure *on highway* and to provide continuity of service by allowing communications (1) in the near neighborhood at direct radio communication range (one-hop neighborhood) and (2) over longer distances (multi-hop neighborhood). For the highway, Table 3.2 indicates that modeling intersections and radio obstacles is not paramount for a minimum realistic mobility model, but that *the models built should at least take into account a vehicle car-following model and a lane change model.* This modeling is based on the choice of a section of highway, whose length is to be determined. The length of the road section must take into account, on the one hand, the maximum ad hoc communication range of the chosen radio technology and the characteristics of the protocol to be studied.

Bands / Technologies and communication modes	Mobile operators	5,9 GHz	Mobile operators and 5.9 GHz
Technologies in Europe	2G, 3G, 4G, soon 5G	ITS-G5 (derived from Wi-Fi)	C-V2X
V2N mode	✓		✓
V2V mode		✓	✓
V2I mode		✓	✓
V2P mode		✓	✓

ITS-G5: *Intelligent Transportation System-G5*[1].

Table 3.3. *Connectivity strategy listed in 2019 (Naik et al. 2019)*

When development of the CBL algorithm began in 2015, the IEEE 802.11p (IEEE 2010) amendment to the IEEE 802.11 (IEEE 2012) standard, specified for vehicular communications, was the only solution that anticipated direct communication between vehicles. It was thus the *IEEE 802.11p* radio technology that was chosen. The first two competing technologies for direct vehicle-to-vehicle (V2V) communications in the 5.9 GHz band (5,875–5,905 Mhz band harmonized by decision 2008/671/EC on the European side and 5,850–5,925 Mhz band in the United States) were as follows: IEEE 802.11p (ITS-G5 or DSRC) technologies and LTE-V2X technology (also called 3GPP C-V2X *Release* 14) (Table 3.3) (Arcep 2019; Molina-Masegosa et al. 2020). The European standard, *Intelligent Transportation System-G5* (ITS-G5), and the American standard, *Dedicated short-range communications* (DSRC), are both based on the IEEE 802.11p standard. LTE-V2X technology, which is newer than ITS-G5 or DSCR technologies and could not be operated with them in 2019, allows, in addition to the use of the cellular network for *vehicle-to-network* (V2N) communications, a direct single-hop communication between two vehicles (Arcep 2019). The development and improvement of solutions for direct and indirect communication between vehicles

1 C-V2X: cellular vehicle-to-everything; V2N: vehicle-to-network; V2V: vehicle-to-vehicle; V2I: vehicle-to-infrastructure; V2P: vehicle-to-pedestrian.

and between vehicles and infrastructure will continue to increase (Singh et al. 2019; Yacheur et al. 2020; Naik et al. 2019; Sehla et al. 2022). Therefore, the new IEEE 802.11bd-2022 standard proposes improvements for V2X communications in order to increase the communication distance and enable higher intervehicular speed than with the current 802.11p standard (Yacheur et al. 2020) (Table 3.4). Similarly, the *5G New Radio (NR) air interface V2X* (5G NR V2X) technology for which the R16 version of C-V2X was released in 2020 is a development of the *Cellular Vehicle-to-Everything* (C-V2X) (Garcia et al. 2021; Sehla et al. 2022; 3GPP 2023). One of the objectives of the IEEE 802.11bd and 5G NR V2X technologies under development will be to guarantee the maximal low latency values expected for certain road safety applications, such as those related to vehicle convoy traffic (maximal latencies of 10–500 ms) or remote vehicle control (maximal latency of 5 ms, Naik et al. 2019).

Characteristics	IEEE 802.11p	P802.11bd
Radio frequency bands	5.9 GHz	5.9 GHz and 60 GHz
Retransmissions	No	Depending on congestion
Countermeasures for the Doppler effect	No	Mid-ambles
Subcarrier spacing	156.25 kHz	312.5 kHz; 156.25 kHz; 78.125 kHz
Supported relative speeds	252 km/h	500 km/h
Spatial flow	One	Multiple

Table 3.4. *IEEE 802.11p standard and P802.11bd draft standard according to Naik et al. (2019)*

With IEEE 802.11p and C-V2X technologies, *direct communications in the 5.9 GHz band have a range of about 500 m for V2V communications* (about a kilometer for V2I communications), relatively low rates and latency times of the order of 10 ms (Arcep 2019), but whose maximum values cannot be guaranteed below 100 ms (Sehla et al. 2022). In order to account for the maximal V2V communication range P of 500 m of the IEEE 802.11p model considered for the study of the CBL scheme (section 3.6.2), *the minimal length chosen for the highway sections used for the definition of highway scenarios (section 3.5) is 5 km (i.e. a minimum* 10×500 m*)*. Running the CBL scheme over a length of at least 5 km will result in obtaining enough occurrences in terms of changes in node state (from leaf to branch and vice versa, Figure 2.26) and modification of the inter-node links (leaf-branch and branch-branch links, Figure 2.29). This will make it possible to collect a number of statistics significantly enough for the computation of structural (section 4.3), routing (section 4.3.2) or application (section 4.4) performance metrics.

3.3. Modeling and simulation tools for VANETs

The modeling of VANETs requires, on the one hand, a mobility model that gives the position of the communication nodes in the network topology at any time and, on the other hand, a communication generator.

The mobility models available have gradually improved over time. Consequently, in the early 2000s, there were only two types of mobility models: traces and synthetic models (the latter are mathematical models) (Camp et al. 2002). As early as 2006, Härri et al. (2006a, 2009) reported two additional tools: on the one hand, survey-based models (surveys that concerned the collection of statistics on the travel behavior of American workers, in particular, by the US Department of Labor); and, on the other hand, models based on traffic simulators, then developed for urban traffic engineering. At the same time, Härri et al. (2006a) and Sommer and Dressler (2008) reported the first studies toward two-way coupling between a traffic simulator and a network simulator. Also, in 2009, Härri et al. (2009) proposed a classification of mobility models (Figure 3.1). These models include models that can provide vehicle traces and that will be the focus of section 3.3.2, namely, stochastic models, trace-based models and models from traffic simulators. PhD students, researchers or engineers who are particularly interested in mobility models for VANETs and stochastic models could start by reading the studies by Härri et al. (2009) and Fiore (2006).

The software tools available for modeling VANETs are also being developed in parallel with the work on mobility models. Three types of software tools used are often identified as: Martinez et al. (2011); Cavalcanti et al. (2018); Ahmed et al. (2019):

– mobility simulators;

– network simulators;

– VANET simulators (Martinez et al. 2011), also referred to as integrated framework by Cavalcanti et al. (2018).

The function of the mobility simulator is to generate realistic vehicle traces, based on the chosen road scenarios. The trace of a vehicle gives a description of the vehicle trajectory, the trajectory being the set of positions taken by the vehicle over time. The generated traces can then be used in network simulators for modeling node mobility. However, the format for describing the trace often differs from that accepted by the network simulator and therefore requires a preliminary conversion. Section 3.3.2 will provide an introduction to different trajectory modeling tools, and in particular based on the work of Fiore (2006), Härri et al. (2009), Rivoirard et al. (2016) and Rivoirard (2018).

Network simulators are used for modeling the communication network. They only model the communication network and consequently require, when used for VANETs, resorting to another tool for retrieving vehicle traces. Network simulators capable of modeling wireless communication now support at least a random mobility scheme such as *random waypoint mobility* (RWP) and the loading of mobility scenarios. Section 3.3.4 will introduce some tools employed for modeling wireless ad hoc communication networks.

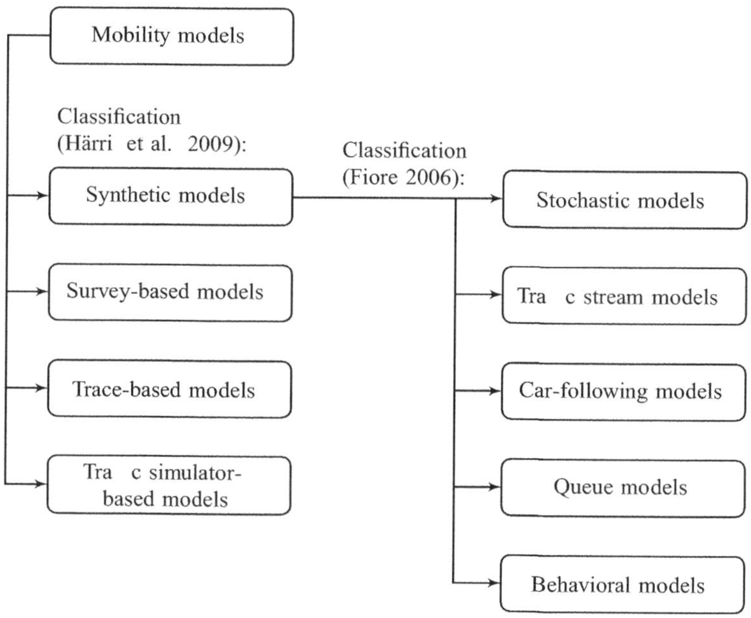

Figure 3.1. *Mobility models for VANETs according to Härri et al. (2009) and Fiore (2006)*

VANET simulators, or integrated frameworks, make it possible for interoperation between a network simulator and a mobility generator. They thus combine different tools into a single development platform that manages their interactions (format, synchronization, etc.). Section 3.3.5 will introduce development environments designed for modeling VANETs.

Finally, Härri et al. (2009) observes that VANET simulators comprise three elements, of which the network simulator and the trace generator are two of the elements; the third being the modeling of the propagation of radio waves in the different simulated geographical environments. Section 3.3.1 introduces some tools for modeling wave propagation.

3.3.1. *Radio wave propagation model*

The radio signal that carries the wireless communication technology information is an electromagnetic wave that interacts with its propagation environment, which makes ground truth difficult to model. The propagated electromagnetic wave undergoes a reflection when it comes across an object (floor, wall and equipment) of very large size compared to its wavelength, and it can then also be refracted; it undergoes diffraction when it meets a surface with sharp contours (such as building corners) on its path between the transmitter and the receiver. Finally, it undergoes dispersion when it comes across small-sized objects in comparison to its wavelength (foliage, road signs, etc.) (Sarkar et al. 2003). The variation over time of the distance between the transmitter and the receiver (a scenario that occurs in V2V communications) also leads to a frequency shift of the wave observed between the transmitting and receiving measurements. This is known as the Doppler effect.

The propagation of the electromagnetic wave is therefore subject to several phenomena. The propagated electromagnetic wave becomes weaker with the distance. This attenuation (*path loss*, where $PathLoss(dB) = 10 \log \frac{P_t}{P_r}$, Figure 3.2) is proportional to the distance between the transmitter and the receiver and is caused by the dispersion of its power in space. It undergoes additional attenuation when it encounters obstacles, called *large scale fading*, which leads to the creation of shadow areas (*shadowing*). These shadow areas are caused, for instance, by signal dispersions on foliage, signal reflections on the ground or buildings and refractions of the signal on buildings. Finally, the power of the wave measured upon its reception shows rapid fluctuations, called small-scale fading, which are due to the delay induced by multipaths and by the Doppler effect (Figure 3.3).

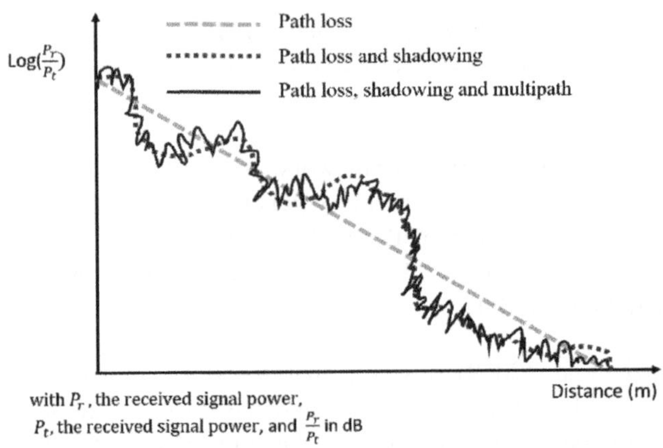

Figure 3.2. *Path loss, shadowing and multipaths of radio waves*

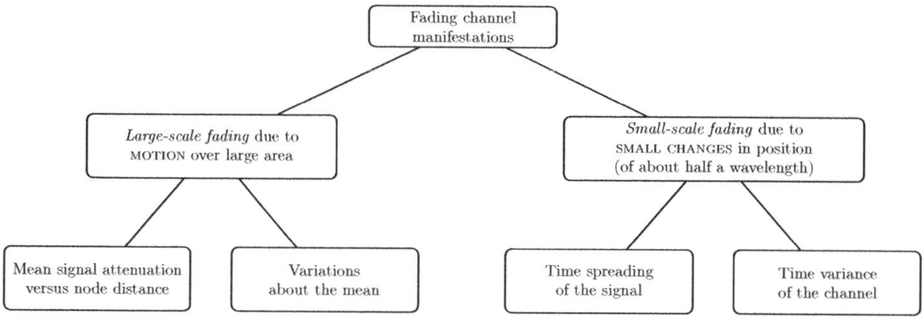

Figure 3.3. *Large-scale and small-scale fading according to Sarkar et al. (2003)*

The various phenomena experienced by the electromagnetic wave during its propagation are responsible for poor signal receptions that have an impact on the operation of ad hoc communication networks. In order to measure the effects on communications, wave propagation models should be used for the purposes of modeling these different phenomena. Dhoutaut et al. (2006), Martinez et al. (2010), Singh and Lego (2011), Rhattoy and Zatni (2012) and Tripp-Barba et al. (2018) show that the choice of a radio wave propagation model has a strong impact on the performance evaluations of routing protocols. In addition, several studies emphasize the need to choose a propagation model adapted to the road environment on which the vehicles operate (highway, urban or rural) (Molisch et al. 2009; Ashtaiwi et al. 2015). In order to reflect this ground truth, which influences the good reception of the transmitted signal, the literature proposes different types of models, namely, deterministic, stochastic models, or even based on measuring electromagnetic wave propagation.

The *free-space path loss model* and the *two-ray ground reflection model* are two examples of deterministic models in which the received power depends solely on distance, and for which the communication area is assumed to be within an ideal circle whose radius is the radio communication range. The *free-space* path loss model, very commonly used for the study of VANETs, simulates the case where both nodes, transmitter and receiver, are in *line of sight* (LOS). The transmitting and receiving antennas are assumed to be in an empty environment; no absorbing obstacles or reflective surfaces are taken into consideration. This model is based on the assumption of a relationship inversely proportional to the square of the distance between the received and transmitted power and considers that below a minimum threshold of receiving power, the receiving node is no longer able to decode the transmitted signals. The equation for this power loss between a transmitter and receiver d away is given by equation [3.1] for the free-space path loss model:

$$Pathloss\_Freespace = \left(\frac{4\pi d}{\lambda}\right)^2 \text{; } \lambda\text{, wavelength of signal} \quad [3.1]$$

The two-ray propagation model considers both the LOS signal and the signal reflected from the ground (Figure 3.4). By considering the multi-path of a transmitted signal, this model takes into account the reflection, refraction and dispersion phenomena of this signal. Mukunthan (2015) recommended the use of the two-radius model for a large distance d between the two communicating nodes ($d > d_c$, where d_c is the critical distance, or first Fresnel zone, beyond which the reception power decreases proportionally to the inverse of d^4). At a short distance, namely ($d < d_c$), for which the decrease in receiving power is much less (inversely proportional to d^2), the *free-space* model is employed. Ren et al. (2017) is an example of how to use this mobility model for the evaluation of a *clustering* algorithm in an urban city scenario, and Amjad and Song (2015) is another example for a highway scenario with multiple intersections.

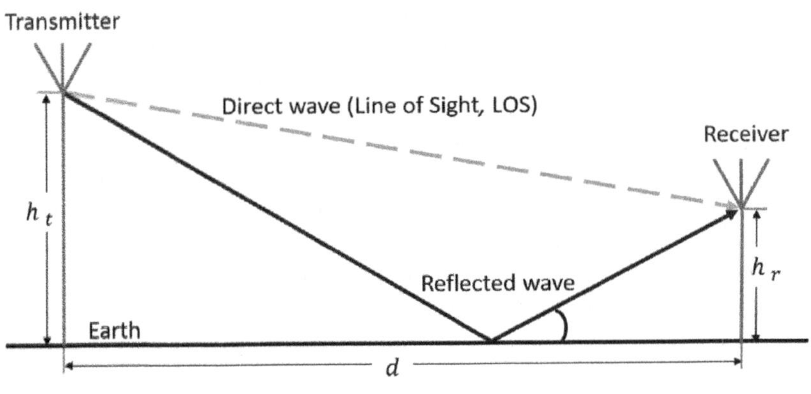

Figure 3.4. *Two-ray ground reflection model*

Probabilistic models include, among others, the *probabilistic shadowing model* (PSM) and small-scale fading models such as the Nakagami model. The probabilistic shadow model takes into account the temporal variability of power by modeling multipaths and fading phenomena using a stochastic approach. This model has two parts. The first part of the model is the mitigation model. This considers the average of the received power with respect to the ratio between the distance of the two nodes (transmitter and receiver) and a reference distance. This ratio is expressed by a power function and a coefficient β whose objective is to determine the rate of decrease of the received power. Greater values of the coefficient β lead to more obstructions in the signal path and a faster decrease. The second part models the changes in the received power at certain distances using a random variable with a Gaussian distribution, zero mean and variance σ. The values of the parameters β and σ are empirically determined by measurements in the real environment. They typically have values of, according to Rhattoy and Zatni (2012), $\beta = 2$ for an open (in free space) environment and $\beta \in [2,7;5]$ for an urban environment, usually with $\sigma \in [4\ dB; 12\ dB]$. Nakagami's model was developed in the 1940s (Sarkar et al. 2003). Its probability density function can be thought of as a generalization of those of other small-scale fading models: a generalization of the Rayleigh model and the Rice or log-normal distributions, according to the chosen parameters (Sarkar et al. 2003; Rhattoy and Zatni 2012). These fading models model the quick fluctuations of the received signals due to the attenuations caused by the transmitted signal multi-paths. The Rayleigh statistical distribution considers that because the transmitted signal is subject to scattering, reflection and dispersion, the transmitted power can reach a hidden area, despite the lack of visibility between transmitter and receiver (*non-line on site* (NLOS)). The Rayleigh density is a reduction of the Rice density, which applies when there is a direct path between the transmitter and receiver (LOS) among the paths taken by the signal. The use of the Nakagami probabilistic model for evaluating the performance of routing protocols is fairly recent (Dar et al. 2010; ETSI 2012; Wu et al. 2013a, 2013b, 2016); it models more realistically wave behavior in real-time mobile communication environments according to Dar et al. (2010).

The reader interested in deterministic and probabilistic models will find their equations in Sarkar et al. (2003) and Rhattoy and Zatni (2012). Viriyasitavat et al. (2015) provided a synthesis of propagation modeling and vehicular communication channels according to road environments (parking lots, highways, intersections, tunnels, urban, rural, etc.). Fallgren et al. (2018) also afforded a summary of the key elements needed to properly set up V2X channels.

Another approach to modeling signal propagation for each environment encountered consists of relying on field testing (Hafeez et al. 2009). Measurement methods and their analysis were proposed in some previous studies (Meireles et al. 2010; Boban et al. 2011; Karedal et al. 2011; Mangel et al. 2011; Abbas et al. 2015). Among these can be found the study by Abbas et al. (2015), which particularly focused on the impact of the shadow effect (*shadowing*) on the propagation of signals caused by neighboring vehicles located between the sender and the receiver. This study was conducted from channel measurements taken in a 200 MHz band around the carrier frequency of 5.6 GHz with a total of 641 frequency points and omnidirectional antennas placed in the center of the vehicle roof, in urban and highway propagation environments and varied traffic densities of southern Sweden. Abbas et al. (2015) characterized V2V links by separating LOS communication situations, those partially or totally obstructed by a vehicle in between (*obstructed line of sight* (OLOS)) and those where the signal is completely blocked by a building (NLOS) making use of GPS data from vehicles and videos taken through windshields. They observed an additional average attenuation of about 10 dB on the received signal strength for OLOS communication. Abbas et al. (2015) then proposed two parameters of a *dual-slope* model, one for urban scenarios and the other for highways, which can be implemented in VANET simulators.

Table 3.5 shows some of the models that have been created to account for the complexity of radio wave propagation phenomena. However, the variety of road configurations (infrastructure, mobility, vehicle density, etc.) is such that it would be necessary to carry out large-scale measurement campaigns to establish relevant propagation profiles for each of them, taking into account the actual traffic observed therein. Besides the issue related to implementing such a measurement campaign, a car fleet that is still not properly connected poses further challenges.

	Analysis from		Propagation context				Environment			Model calibrated with measurements				
Models	Model	Measurement	LOS	NLOS	OLOS	*Fading*	Highway	Urban	Rural	*Free space*	*Two Rays*	PSM	Nakagami	Another one
Conventional models														
Free space	✓		✓				✓		✓					
Two rays	✓		✓	✓			✓		✓					
PSM	✓		✓	✓		✓	✓		✓					
Nakagami	✓		✓	✓		✓								
Models for VANET														
(Mangel et al. 2011)		✓		✓				✓						
(Meireles et al. 2010)		✓	✓	✓	✓		✓	✓						
(Boban et al. 2011)	✓	✓	✓		✓	✓	✓							H
(Karedal et al. 2011)	✓	✓	✓	✓		✓	✓	✓	✓		R	HU		
(Abbas et al. 2015)	✓	✓	✓		✓	✓	✓	✓					HUR	

LOS: line of sight, NLOS: non-line of sight, OLOS: obstructed line of sight, PSM: probabilistic shadowing model, H: highway, U: urban, R: rural.

Table 3.5. *Some models of signal propagation in VANETs (Rivoirard 2018)*

3.3.2. Models for obtaining vehicle traces

3.3.2.1. Real trace databases

Vehicle traces can be constructed from databases of actual vehicle traces. These data are usually extracted from global positioning system (GPS) data records or video recordings.

An example of a database is the *Community Resource for Archiving Wireless Data At Dartmouth* (CRAWDAD) (Crawdad 2020a). This public database allows the scientific community to access more than a hundred sets of GPS traces of taxis and buses. Many works in the literature make use of this database (Crawdad 2020b), such as those of Li et al. (2013), which show the temporal limitation of V2V connections induced by the high mobility of the nodes of a VANET. Nonetheless, the mobility of a fleet of taxis or buses is constrained by its professional purpose. In addition, these fleets concern only a subset of road traffic and therefore cannot be representative of the mobility of road vehicles, which generally account for greater numbers on the road. There is thus a lack of trace data for private vehicles with which these fleets interact. As a result, the actual traces originating from these records do not contain vehicles in close proximity to each other or include very few vehicles.

With databases created from video recordings, two large groups of datasets have been built since the 2000s according to Buisson et al. (2016): the Delft University of Technology, Technische Universiteit Delft (TU Delft) (Hoogendoorn et al. 2003), which is non-public, and the traffic analysis tools *Next Generation Simulation* (NGSim) (FHWA 2020) of the U.S. Federal Highway Administration program (FHWA). NGSim data are public but unfiltered, but filtering methods are proposed in the literature, such as that of Ferreira et al. (2009). In addition, they cover a road section of only 1 km, which is too short a length to be representative of a section of highway road. A third set of filtered video data were proposed by Buisson et al. (2016). These were data from aerial video recordings, from a helicopter, in the Grenoble region over an expressway area of 500 m. Recorded for purpose of the PREDIT MOCoPo project, these filtered data can be utilized to reconstruct the trajectories of the vehicles in motion within this area. The set of resulting traces, used by Rivoirard et al. (2017) for the comparative evaluation of the performance of the *Optimized Link State Routing* (OLSR), AODV, GRP and DSR routing protocols, has the advantage of containing the trajectories of all the vehicles that have crossed the section covered by the recording and of showing fluidity and congestion phenomena specific to road traffic (Figure 3.5): three groups of trajectories represent, respectively, from left to right, fluid traffic; congested traffic, slowdown and "stop & go" phenomena (frequent stops and starts); finally, a hybrid traffic situation, partly fluid, partly congested. However, with only 500 m of road length covered, these data do not allow the evaluation of a new routing protocol for VANETs designed for highway sections.

3.3.2.2. *Stochastic models*

Stochastic mobility models are derived from the *Random Waypoint* (RWP) model that was originally designed by Johnson and Maltz (1996) for the evaluation of the *Dynamic Source Routing* (DSR) protocol developed for mobile ad hoc networks (MANETs). In the RWP model, each node is initially placed at a randomly chosen position in a plane simulation space whose dimension is previously predefined. During the simulation, each node stops at its current position (*waitpoint*) for a pause time and chooses a new position towards which it moves in a straight line, at a constant speed randomly drawn within a given range of values. Throughout the simulation, the nodes will alternately move from their current position to their next destination, and then pause once that position is reached. The resulting traces are those of zigzag trajectories. Although unrealistic, this mobility model of easy implementation has been used in many protocol evaluations (Sondi et al. 2010; Tian et al. 2010; Daeinabi et al. 2011; Mohapatra and Kanungo 2012; Roy et al. 2015; Cha et al. 2016). It is a model that is conventionally supported by network simulators adapted to MANET or VANET.

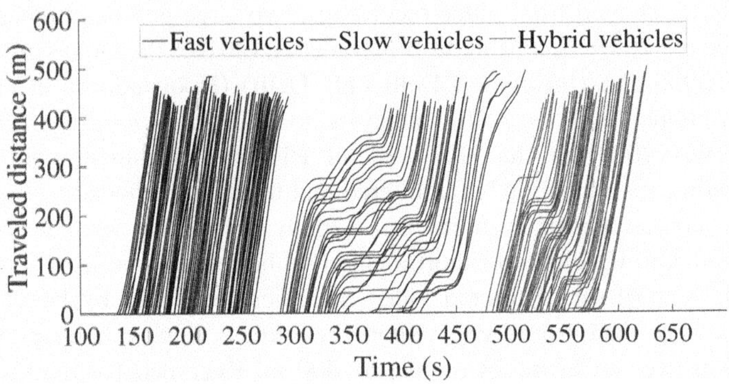

Figure 3.5. *Trajectories traveled over 500 m of the RN97 (southern suburb of Grenoble) in September 2011, of three groups of vehicles, with respect to time (MOCoPo data) (Rivoirard et al. 2016). For a color version of this figure, see www.iste.co.uk/wahl/design.zip*

Stochastic models specifically defined for VANETs have mobility descriptions for which the random motions of the nodes are constrained in a graph that represents a road topology (Fiore 2006). The movements of the nodes (vehicles) can be individually generated or by vehicle group. For this purpose, the graph utilized can take various forms. It can be a graph randomly generated by a model (for example,

Manhattan model or Voronoi diagram), obtained from extracting the road topology from a real map using topological standards such as GDF, which has recently been covered by a standard for ITS (*ISO 20524-1:2020 2020 Intelligent transport systems – Geographic Data Files (GDF) GDF5.1 – Part 1: Application independent map data shared between multiple sources GDF* standard) or map data in the TIGER (*Topologically Integrated Geographic Encoding and Referencing*) format (Härri et al. 2009), or still a user-defined graph. Figure 3.6 illustrates examples of such maps found in the literature.

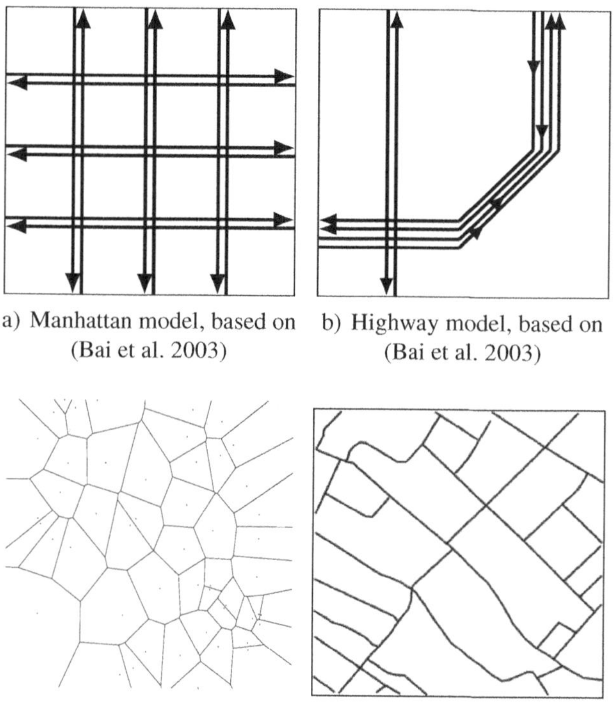

a) Manhattan model, based on (Bai et al. 2003)

b) Highway model, based on (Bai et al. 2003)

c) Voronoi diagram (generated with MATLAB)

d) Model extracted from a real map in GDF format, based on (Fiore et al. 2007)

Figure 3.6. *Sample maps. For a color version of this figure, see www.iste.co.uk/wahl/design.zip*

Various improvements have been made to the RWP model to better approximate road traffic dynamics. For example, in the *Street random waypoint* (STRAW) urban mobility model – which is utilized for evaluating the performance of *Optimal Next Hop Selection* (OHNS) (Lin and Huang 2012), DSDV (He 2002) and *QoS Optimized Link State Routing* (QOLSR) (Sondi et al. 2013) protocols – the map paths followed by the nodes between two successive positions are either selected by a shortest path algorithm (such as Dijkstra-based), or stochastically by introducing a probability of change of direction (turn-right, turn-left) at each intersection encountered. However, the *street random waypoint* model does not consider interactions between vehicles, and these can then collide with each other. The *freeway mobility model* (FDM) of Bai et al. (2003), used by Wu et al. (2013a, 2013b, 2016), employs as a road map that of a two-lane highway (Figure 3.6b) and adds to the *street random waypoint* model the notion of interaction between vehicles that was missing: each node that travels on the highway maintains a minimum safety distance from the node preceding it. Node speed at a given time depends both on its speed at the previous time and on a variable whose value is randomly drawn, but its motion is limited by the minimum safety distance that separates it from the node preceding it on the same lane. These two examples show that despite the contextual improvements made to the original RWP model, stochastic models struggle to reflect the reality on the ground (Rivoirard 2018). They are constructed from randomly selected positions and speeds, and the significant dispersion of velocity values over time and space, which is an important feature of VANET networks, is not taken into account.

3.3.2.3. *Traffic simulators and trace generation*

Synthetic models are employed by traffic simulators whose purposes vary, depending on whether they are macroscopic, microscopic or mesoscopic.

Microscopic modeling is concerned with the individual behavior of each of the vehicles in the road network deemed to be interacting with other vehicles and the road infrastructure. It makes use of parameters such as distance between vehicles (Härri et al. 2009), speed of travel, acceleration, deceleration, as well as overtaking criteria, behavior at road intersections and in the presence of signals, human behavior according to different attributes (age range, etc.). Using car-following models, microscopic simulators generate the trajectory of each vehicle using patterns of interaction between vehicles that are in their close vicinity. Therefore, *microscopic simulators are interesting tools for the generation of trajectories necessary for modeling node mobility in a VANET* (Martinez et al. 2011; Patel 2012; Ahmed et al. 2019). Nevertheless, they are memory-intensive and result in long-running simulations. As such, they need to restrict simulations to limited portions of the road network (a few streets) (Rivoirard 2018).

Macroscopic simulators, whose implementation can be fast and suitable for large-scale simulations (the size of a city), use traffic flow models. In this way, macroscopic modeling of road traffic is similar to that of a hydrodynamic flow, section by section. It may concern the topology of the road, road density, the constrained movements of vehicles, the speed limits at road level, the number of lanes, overtaking and road safety rules and intersection signs (Martinez et al. 2011).

Macroscopic simulators are suitable for the analysis of road traffic on a macroscopic scale. They do not make the generation of individual vehicle trajectories possible.

Mesoscopic simulators employ queuing models (Fiore 2006) and combine the properties of macroscopic and microscopic models, considering individual vehicle mobility, but adopting a macroscopic approach to their movements and an average speed between two travel points (Martinez et al. 2011). Mesoscopic simulators therefore take traffic into consideration as homogeneous vehicle platoons (Ahmed et al. 2019) and do not allow the generation of individual trajectories of independent vehicles.

3.3.3. *Mobility simulators*

Given that stochastic models are not realistic enough to simulate vehicle mobility (section 3.3.2.2) and that actual traces still contain an insufficient number of nearby vehicles that could likely interconnect as part of a network (section 3.3.2.1), it is possible to compensate for these shortcomings by constructing traces. This construction can be done on the basis of a realistic mobility scenario, but also on the basis of data on the usual traffic conditions in certain areas. For example, Sondi et al. (2016) uses traces reconstructed from Google Maps data in urban areas. This reconstruction can be done by means of a microscopic simulator that automatically reconstructs traces for high densities.

Many microscopic mobility generators are referenced in research reviews on VANETs or vehicular networks. Ahmed et al. (2019) presented 14 of them, some available under free licenses and others under commercial licenses. Cavalcanti et al. (2018) retained only three of these in their study relating to the research carried out on VANETs and freely licensed tools used between 2010 and 2018: first, the two traffic generators *Simulation of Urban MObility* (SUMO) (Lopez et al. 2018) and *Vehicular Ad Hoc Networks Mobility Simulator* (VanetMobiSim), which are the most commonly used with 31.2% and 11.8%, respectively, of use in the literature; then the CityMob simulator (Martinez et al. 2008; GRC 2009) is created to easily generate urban scenarios from three mobility models (*Simple* model for vertical and horizontal mobility without change of direction, Manhattan (Figure 3.6(a)), and *Downtown*, a model that adds the notion of different traffic densities with areas where vehicles drive more slowly). The characteristics of the simulators that we found most referenced before 2018 for protocol routing applications are summarized in Table 3.6. In these tools, SUMO is the most widely used simulator in routing protocol studies for VANETs (Rawashdeh and Mahmud 2012; Arkian et al. 2015; Azizian et al. 2016; Ren et al. 2017; Sethi and Chand 2017), followed by VanetMobiSim (Bitam and Mellouk 2011; Wahab et al. 2013; Zheng et al. 2014; Li et al. 2015; Sun et al. 2015).

Microscopic simulators	VanetMobiSim	TSIS-CORSIM	SUMO	NetStream
References	(Härri et al. 2006b)	(Halati et al. 1997)	(Krajzewicz et al. 2010)	(Teramoto et al. 1997)
Software				
First version	2006	1997	2002	1997
Current Version	2007	2012	2021	1997
In development		✓	✓	
Cross-platform			✓	
Free License	✓		✓	
GUI	✓	✓	✓	✓
Ease of Use	+++	+	++	–
Road network representation				
Random map	✓		✓	✓
User-defined map	✓	✓	✓	
Imported real map	✓	✓	✓	✓
Modeled functions				
Car-following model	✓	✓	✓	✓
Multi-lane	✓	✓	✓	
Overtaking	✓	✓	✓	
Speed constraint	✓	✓	✓	
Intersection management	✓	✓	✓	
Different vehicle types	✓	✓	✓	

Table 3.6. *Some microscopic road traffic simulators (Rivoirard 2018)*

The SUMO simulator for urban mobility (Krajzewicz et al. 2010) enables modeling intermodal traffic systems, including road vehicles, public transport and pedestrians. The program can import the road network manually, randomly or operate from a real road network. SUMO is able to manage multiple road lanes, maximum speeds per lane, as well as intersections through a stop or traffic light system. This simulator can also be used to model several types of vehicles. It offers different car-following algorithms, for example, the Krauß algorithm (Krauß 1998), which describes the motion of each vehicle according to the positions and speeds of nearby vehicles. In this model, each vehicle tries to reach its "desired speed" while maintaining a perfect safe distance from the vehicle ahead. This freely licensed simulator is coded in C++. It can be configured using XML (*Extensible Markup Language*) files, as is the VanetMobiSim simulator. Since 2007, the extension MOVE (*Mobility model generator for vehicular network*) for SUMO, which is a Java application on top of SUMO (Lan 2009; Mantoro and Reza 2016), has allowed the generation of vehicle traces compatible with communication network simulation software such as Ns-2, QualNet or GloMoSim, according to Ahmed et al. (2019).

3.3.4. *Communication network simulators*

Various articles (Martinez et al. 2011; Al-Sultan et al. 2014) proposed in-depth comparative studies of communication network simulators used for communication networks, in particular, for MANETs and VANETs. Network simulators are tools whose use and capabilities are continuously developed over time. Therefore, before choosing a tool, it is advisable to check the dates of these studies and consult the latest development information on the respective websites of the simulators. In addition, a communication system is classically defined on a set of protocols developed on mainly the five layers of TCP/IP stack (according to the terminology of the OSI ISO system, which defines seven layers): the physical, data link (including the *Medium Access Control* (MAC) sublayer), network, transport and application layers. However, the utilization of a simulator should not require users to program protocols on each of the layers. A set of protocol algorithms is expected to be already implemented in the simulator upon which users can rely in order to develop only the studies that concern them. For example, developers of ITS applications might wish to be able to reuse layers one to four protocols, which would be of interest to ITS, on the one hand, and already implemented in the simulator, on the other hand. It would be required in particular that the simulator provides support for node mobility, namely to be able to parameterize the protocols of the layers relating to the radio technology (or technologies) used (physical and data link layers), as well as those necessary for ITS applications at the level of network and transport layers. On the other hand, redeveloping these different protocols should not be the responsibility of developers. Furthermore, because they are often originally designed for research and development purposes, some simulation tools are free software, while others become commercial software programs. The cost of the simulator (and especially its maintenance) can then be used as a choice criterion.

Simulators	NS-2	NS-3	MATLAB	Omnet++	OPNET Riverbed Modeler	QualNet
References	(nsnam 2014)	(nsnam 2021)	(MathWorks 2021)	(OpenSimLtd 2019b)	(Riverbed 2021)	(Scalable 2021)
Preference in the literature in % (Cavalcanti et al. 2018)	37.9%	7.1%	8.6%	8.1%	4.5%	1.5%
Software						
Development in progress	×	✓	✓	✓	✓	✓
Simulation language	C++, OTCL	C++, Python	C, C++, Java	C++	C++, Java	Parsec C++
Free, academic or commercial license	L	L	C	A and C	C	C
GUI	✓	✓	✓	✓	✓	✓
Animation	✓	✓	✓	✓	✓	✓
Supported function						
Scaling	–	–	++	+	+	++
802.11p model	✓	✓	*WLAN Toolbox*	✓	✓	✓
LTE model	×	✓	*LTE Toolbox*	SimuLTE	*LTE License*	✓

Table 3.7. *Some network simulators*

Table 3.7 shows a few characteristics of freely licensed or commercially licensed network simulators with their percentage of use calculated by Cavalcanti et al. (2018), based on the analysis of 198 articles on VANETs. Network Simulator (ns) and Objective Modular Network Testbed in C++ (OMNeT++) are examples of discrete event simulators freely licensed or that offer a free academic license.

Network Simulator (ns) is a software program developed for research studies on communication networks. There are currently two existing distinct versions, ns-2 (nsnam 2014), developed from 1995 to November 2011 (latest version ns-2.35), and ns-3 (nsnam 2021), developed since 2006 (January 2021 version: ns-3.33). The architecture of the ns core, previously developed in C++ or with OTcl scripts (ns-2), has been revised for the ns-3 version: the kernel has since been developed in C++ or with Python scripts. More recently, ns-3 appears in 2018 in fewer scientific publications on VANETs (Cavalcanti et al. 2018). Nonetheless, with the end of the development of the ns-2 program kernel to give priority to those of ns-3, ns-2 will gradually become obsolete. The ns-3 library proposes many models for modeling a wireless communication system, including Internet models (TCP, UDP, IP including IPV6, etc.), routing protocol models of DSR, DSDV, AODV, OLSR protocols, *Wireless Access in Vehicular Environments* (WAVE) models, including an 802.11p model, a *Long Term Evolution* (LTE) module and physical layer modules (antenna, propagation and spectrum).

The *Objective Modular Network Testbed in C++* (OMNeT++) discrete event simulator provides an integrated development and simulation environment, developed in C++ and based on the Eclipse platform (OpenSimLtd 2019b). The Eclipse platform comprises many modules developed under a free license (GPL or LGPL) by researchers. For the design and modeling of new communication protocols, the *INET Framework* library can be used. It is an *open-source* library of models for the OMNeT++ simulation environment that supports many communication networks (wired, wireless, mobile, ad hoc and sensor networks). It includes the Internet Protocol stack (TCP, UDP, IPv4, IPv6, etc.), improves support for the wireless physical layer, routing protocols for MANETs, and provides support for node mobility (OpenSimLtd 2019a, 2019c).

MATrix LABoratory (MATLAB) is one of the most widely used commercially licensed software programs. It is both a development environment and an emulated programming language for data analysis, algorithm development and model creation. This is not a network simulator, but propositions, such as the *VANET Toolbox* of Wang et al. (2019); Wang (2020), which supports V2V communications with lane change and car following, have been developed and proposed by contributors. Earlier examples of uses for the evaluation of *clustering* algorithms for VANETs include those of the *Vehicular clustering based on the Weighted Clustering Algorithm* (VWCA) solution by Daeinabi et al. (2011), *Particle Swarm Optimization Cluster* (PSO-C) by Harrabi et al. (2016) or *Quality-of-service-based clustering protocol for vehicular ad hoc networks* (VANET QoS-OLSR) by Wahab et al. (2013).

Less referenced in the literature due to their licensing costs, the *Riverbed Modeler* and Qualnet simulators are commercial software originally developed by research organizations.

The *Riverbed Modeler* (Riverbed 2021) simulator, formerly known as *Optimized Network Engineering Tools Modeler* (OPNET Modeler), is a commercial software for the design, modeling and analysis of communication networks that offers models validated by the software publisher of the most widely used protocols (OLSR, AODV, TCP, UDP, IP, Ethernet, IEEE 802.11, etc.); and, since version 18, provides modeling of the 802.11p interface. It also gives the possibility to manually generate node trajectories or to use random motion models. The *Wireless Modeler suite* license required for the development of wireless networks can be supplemented with additional licenses, for instance, those of protocols under development, such as LTE.

Examples of academic research on routing protocols include Sondi et al. (2010) for the design and evaluation of the QOLSR, Hnatyshin and Asenov (2010) for the development of the *Geographical Ad hoc On-demand Distance Vector* (GeoAODV) protocol and Hussey et al. (2013) for a comparison of proactive and reactive geographic routing protocols, or still the research of Rivoirard et al. (2018); Wahl et al. (2021) related to the works on the *Chain-Branch-Leaf* OLSR (CBL-OLSR) protocol.

Qualnet (Scalable 2021) was originally built from the *Global Mobile Simulator* (GloMoSim) open source license by Scalable Network Technologies, Inc. in 1999. GloMoSim is a large-scale simulation software library for wireless and wired networks (Zeng et al. 1998) and Qualnet is a network simulation tool specified for business or mission-critical information and communication systems. Developed in Parsec C++ (*Parallel Simulation Environment for Complex Systems*), this tool provides solutions for the design, testing and analysis of communication systems for different application domains, such as automotive, marine, airborne army and critical infrastructure. It also proposes solutions for integrating real or simulated cyber-attacks on simulated wired and wireless networks. Kumar et al. (2016) is a usage example for benchmarking AODV, DSR, *Dynamic MANET on demand Routing* and *Zone Routing Protocol* (ZRP) routing protocols in ad hoc networks with constant traffic load and evaluation based on performance metrics such as jitter, throughput, mean end-to-end delay and packet delivery ratio.

3.3.5. *VANET simulators*

A VANET simulator consists of at least two simulators: one for modeling communication networks and the other for modeling mobile node trajectories. There are various VANET simulators available. Weber et al. (2021) offer a recent, detailed overview, focusing on their use with respect to their ability to model new technologies – such as 5G, *Software-defined network* (SDN) or *edge computing* – and, on the other hand, to provide support for dependability and security mechanisms. Some examples of VANET simulators are shown in Table 3.8.

Simulators	VEINS	VENTOS	VsimTRI	NCTUns	EstiNet
References	(Sommer et al. 2011)	(Amoozadeh et al. 2019)	(Protzmann et al. 2017)	(Wang and Lin 2008)	(EstiNet 2020)
Characteristics					
First version	2006	2013	2010	2002	2011
Development in progress	✓	✓	✓	–	✓
Mobility generator	SUMO	SUMO	SUMO, VISSIM, PHABMACS	Embedded	Embedded
Network simulator	OMNeT++	OMNet++	OMNeT++, SNS, NS-3, SWANS	NCTUns	Estinet
Traffic model	Mic*	Mic and Mes*	Mic	Mic	Mic
Cellular network simulation supported			✓		
Features					
Car-following model	✓	✓	✓	✓	✓
Traffic light	✓	✓	✓	✓	✓
Multi-stream	✓			✓	✓
Speed model	✓		✓	✓	✓
Dynamic traffic routing			✓		

* Mic: microscopic; Mes: mesoscopic.

Table 3.8. *VANET simulators (Ahmed et al. 2019; Amoozadeh et al. 2019)*

VEhicle In Network Simulation (VEINS) is an open-source VEC simulation tool (Sommer 2020; Sommer et al. 2011, 2019) that integrates OMNeT++ with SUMO.

VEhicular NeTwork Open Simulator (VENTOS) was designed to study vehicle traffic flows, collaborative driving and interactions between vehicles and the infrastructure via DSRC wireless communications (Amoozadeh 2017; Amoozadeh et al. 2015). It is a platform written in C++ that integrates SUMO and OMNeT++.

EstiNet Network Simulator and Emulator (EstiNet 2020) is the commercial version created in 2011 based on the *National Chiao Tung University network simulator* (NCTUns) tool (Wang 2013). It is a simulator that embeds both a proprietary network simulator and its own mobility simulator. In particular, NCTUns was used for the evaluation of the C-DRIVE protocol (Maslekar et al. 2011). A first version of the

802.11p interface was developed in 2013 as part of the commercial version EstiNet by Wang et al. (2013).

More recently, "Eclipse MOSAIC" (DCAITI 2021) is the name with which the free license of the *Vehicle-2-X Simulation Runtime Infrastructure* (VSimRTI) software has been distributed since October 2020. Its last version, at the time of writing, Eclipse MOSAIC 21.1, was released in October 2021. VSimRTI was developed by the Daimler Center for Automotive Information Technology Innovations (DCAITI) at the University of Technology in Berlin (Protzmann et al. 2017). It offers a complete cosimulation environment for the evaluation of new solutions for cooperative ITSs (Omnetpp 2010; DCAITI 2021). Eclipse MOSAIC is a runtime environment that pairs different simulators: the SUMO and the sub-microscopic scale *PHysics Aware Behavior Modeling Advanced Car Simulator* (PHABMACS) traffic simulators (Fraunhofer FOKUS 2020); ns-3, OMNeT++, *Simple Network Simulator* (MOSAIC SNS) (Eclipse 2021) communication simulators dedicated to the transmission of V2X messages using ad hoc communications and MOSAIC Cell for the simulation of cellular communications; an application simulator; as well as various visualization and analysis tools.

3.4. Choice of modeling and simulation tools

Different criteria can be employed in the choice of software tools such as cost (acquisition cost, development time, etc.), model availability, the possibility of extending these models or developing others, accessible documentation, efficient IT support, user forums and local know-how. Our selection criteria were the result of an expertise previously developed on simulation tools and their availability at the development site.

For the system modeling of the CBL scheme, the OPNET network simulator *Riverbed Modeler* (Riverbed 2021), as well as the MATLAB development environment were used. By making the most of the protocol models developed and validated by the publisher of *Riverved Modeler* for the system model of CBL, we were able to concentrate all our efforts on the modeling and integration of the CBL scheme into an ad hoc wireless communication station model of the *Riverbed Modeler* platform. MATLAB was first used for the structural modeling of CBL and for the protocol debugging of the CBL scheme prior to its implementation under *Riverbed Modeler*. It was then employed for the validation of the *Riverbed Modeler* implementation of the CBL scheme and structural performance metrics by comparing the results obtained on the CBL structure through the two simulators.

For the creation of the trajectories necessary for node mobility, we resorted to the free traffic generation software SUMO and the MATLAB suite, the latter being used here for the conversion of the trajectories obtained in SUMO into the format accepted by the OPNET simulator.

3.5. Definition of mobility scenarios in a highway context: example of the construction of mobility models

The open-source software SUMO was chosen as a microscopic simulator for the generation of motorway scenarios because of its frequent use by the scientific community working on research related to ad hoc networks. This tool was able to generate two different types of road networks in order to obtain significantly large VANETs with configurations where infrastructure is likely to be non-existent: highways and roads outside urban areas.

The first three road networks correspond to the following conventional motorway configurations (Figure 3.7):

– R1, a 5 km long three-lane section of a one-way highway;

– R2, a 5 km long two-way section of a three-lane highway;

– R3, a 5 km section of highway, with two times three lanes with a highway entrance and exit. The exit is located 1.8 km from the start of the section, while the entrance is 3 km away.

OSI layer no.	Evaluation / Protocol	Matlab (Str)	*Riberbed Modeler*, RM (Sys)
7	Message and application traffic	–	✓
4	TCP/UDP stack	–	*RM*'s native code
3	IP suite	–	*RM*'s native code
3	CBL	CBL scheme	–
3	CBL-OLSR	–	CBL embedded in *RM*'s native OLSR code
1 and 2	IEEE 802.11p radio interface	–	*RM version 18.1* native code

Table 3.9. *Protocol layer modeling 1 to 7 according to the analysis performed*

Evaluation / Highway	Matlab (Str)	*Riberbed Modeler,* RM (Sys)
Mobile nodes	SUMO trajectories	SUMO trajectories converted to ".trj"
Wave propagation		
Profile 1: LOS	Communication distance of 500 m	*RM* settings (*Freespace* and communication distance of 500 m)
Profile 2: LOS/OLOS	✓	*RM* settings + LOS/OLOS profile to be implemented

Table 3.10. *Modeling of terrain characteristics according to the analysis performed*

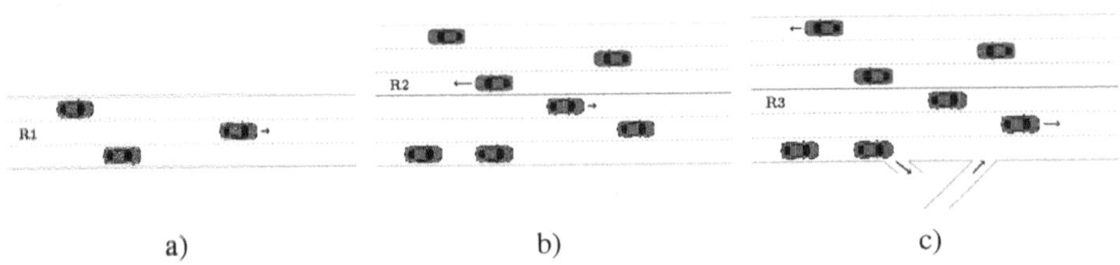

Figure 3.7. *Motorway networks: (a) R1, (b) R2 and (c) R3*

Two additional road networks were built from the actual A27 highway network (Figure 3.8):

– R4 consists of a 5 km section of the A27 highway which connects the French city of Lille to Belgium and a 1 km section of the RD90 secondary road. The A27 is a dual carriageway. The RD90 is a two-lane road running perpendicularly through the A27 section being studied.

– R5 contains only the section of the A27 motorway.

The *creation of a road network* with SUMO can be done manually (Figure 3.9(a)) or by importing a real network exported from OpenStreetMap (Open StreetMap 2021) (Figure 3.9(b)). Road networks R1, R2 and R3 were created in manual mode. To this end, it is possible to directly write the necessary configuration files or to use the *netedit* graphical network editor in SUMO (Figure 3.9(a)), which, built on *netconvert*, makes it possible to create and modify road networks for SUMO (DLR 2021b). *netconvert* (DLR 2021a) is a command line that imports and converts road networks from different sources to create one that can be used by SUMO tools.

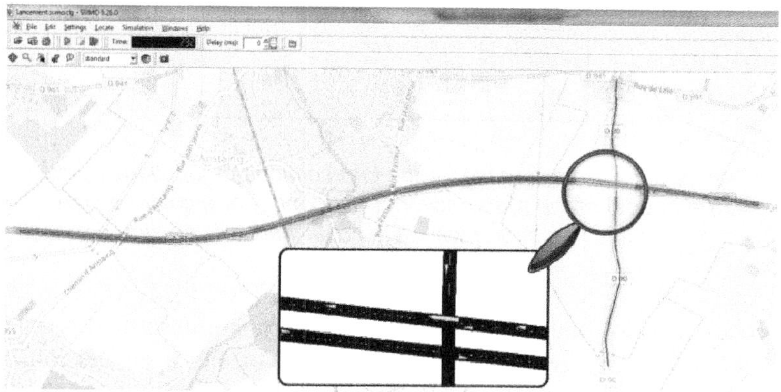

Figure 3.8. *A27 highway and RD90 road imported from Open Street Map into netedit using the JOSM tool and resulting R4 network in SUMO with road traffic (S10 scenario) (Rivoirard 2018). For a color version of this figure, see www.iste.co.uk/wahl/design.zip*

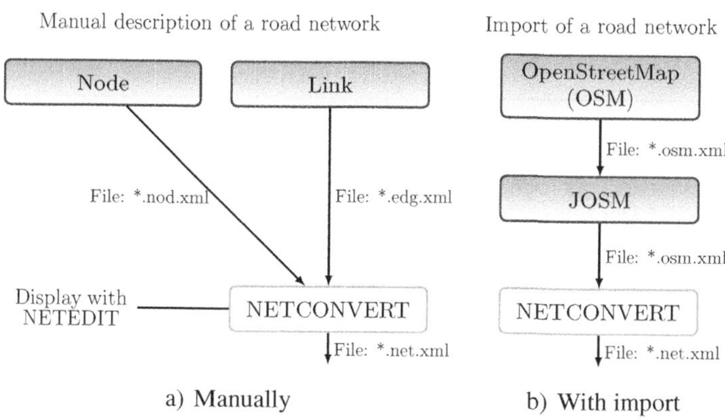

Figure 3.9. *Construction of a road network for SUMO (Rivoirard 2018)*

In manual mode, road network nodes, which correspond to road intersections or junctions, were first defined by their geographic coordinates and unique identifier in a file with the ".nod.xml" extension, as shown in Figure 3.10.

Then, the links between the nodes of the road network were specified in a file with the ".edg.xml" extension in which the number of lanes and the maximum authorized speed can be specified (Figure 3.11), but also a priority to the road section, a width, a traffic authorization dedicated to specific vehicles.

```
<nodes>
  <node id="1" x="0" y="0" />
  <node id="2" x="5000" y="0" />
</nodes>
```

Figure 3.10. *Definition file for the intersections or junctions points (the nodes) of the road network (.nod.xml). For a color version of this figure, see www.iste.co.uk/wahl/design.zip*

The creation of R4 and R5 was carried out by importing a real road network (Figure 3.9(b)). The cartographic data used are those of the "open and collaborative map of the World" project, under free licenses, *OpenStreetMap* (OSM) (OpenStreetMap 2021). The JOSM (Java OSM) editor of *OpenStreetMap* was used to edit the OSM map data and to select only those relating to the A27 highway (R4 and R5 networks) and the RD90 secondary road (R4 network). The resulting map data, in .oms.xml format, is then opened directly with the *netedit* interface (Figure 3.8).

```
<edges>
  <edge from="1" id="1to2" to="2" numLanes="3" speed="36.1"/>
  <edge from="2" id="2to1" to="1" numLanes="3" speed="36.1"/>
</edges>
```

Figure 3.11. *Script for defining connections (that is roads or streets) between junctions points in the road network (.edg.xml). For a color version of this figure, see www.iste.co.uk/wahl/design.zip*

The *road network generation* in the ".net.xml" extension format accepted as input by SUMO is done using the *netconvert* command line, which operates the import and conversion of the ".nod.xml" and ".edg.xml" files manually defined (Figures 3.9(a) and 3.12), or the ".osm.xml" file defined by import (Figure 3.9(b)).

The next step consists of the *definition of the demand for vehicles* by way of a road file with a ".rou.xml" extension (Figure 3.13). This file specifies the set of vehicle types and the demand for vehicles for each of the flows defined on routes consisting of links between nodes of the road network (Figure 3.16). Vehicle types are characterized by their size, maximum acceleration and deceleration, and even maximum speed. The combination of a traffic request file and a road network on input in SUMO will then result in generating traces (Figure 3.13).

In order to define the demand for vehicles on the R1 to R5 networks, different traffic densities were considered. Three scenarios of traffic densities were first defined for the R1, R2 and R3 road networks: low, medium and high. Each of these densities had a ratio of one-sixth of trucks and five-sixths of cars. For the R3 network, the following constraint was imposed: 25% of vehicles come in through the motorway entrance, 25% of vehicles leave through the exit and the remaining 50% vehicles complete the whole motorway section. Nine scenarios, denoted S1 to S9, were thus specified, whose characteristics in terms of density and road network are shown in Table 3.11.

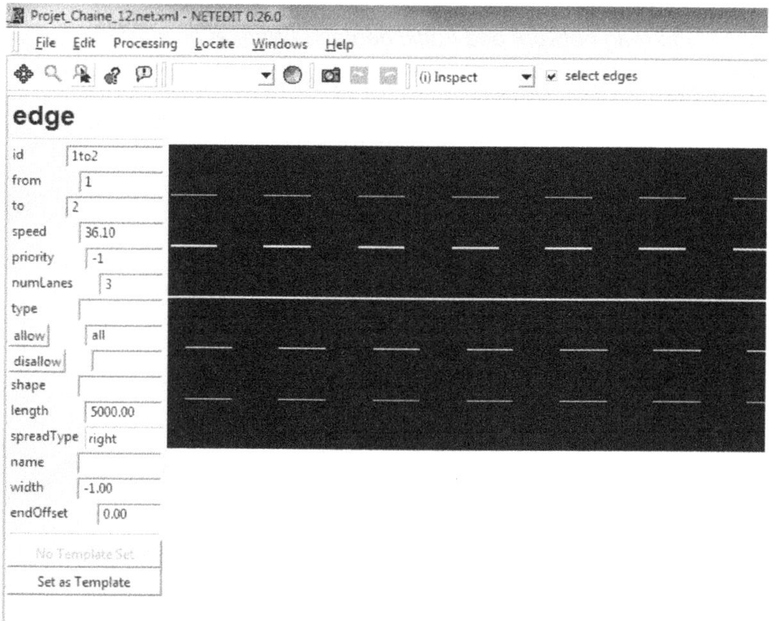

Figure 3.12. *Visualization in netedit of the R2 network created by netconvert (Rivoirard 2018). For a color version of this figure, see www.iste.co.uk/wahl/design.zip*

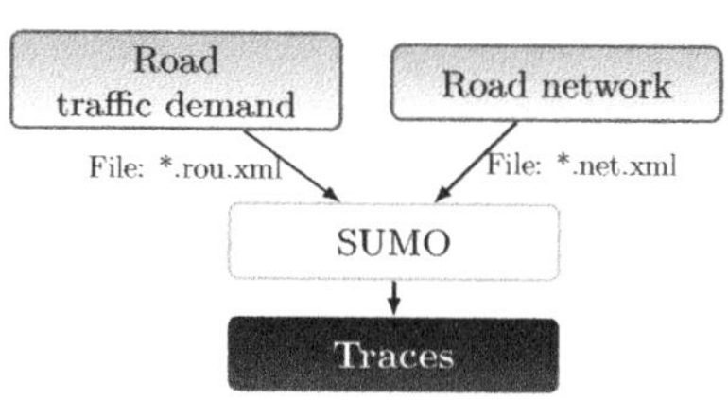

Figure 3.13. *Constructing traces with SUMO (Rivoirard 2018)*

Density	Car Traffic *(veh/h/direction)*	Truck traffic *(veh/h/direction)*	R1	R2	R3
Low	500	100	S1	S4	S7
Medium	2,000	400	S2	S5	S8
High	4,000	800	S3	S6	S9

Table 3.11. *Definition of scenarios (S1 to S9) according to road network and traffic demand (Wahl et al. 2021)*

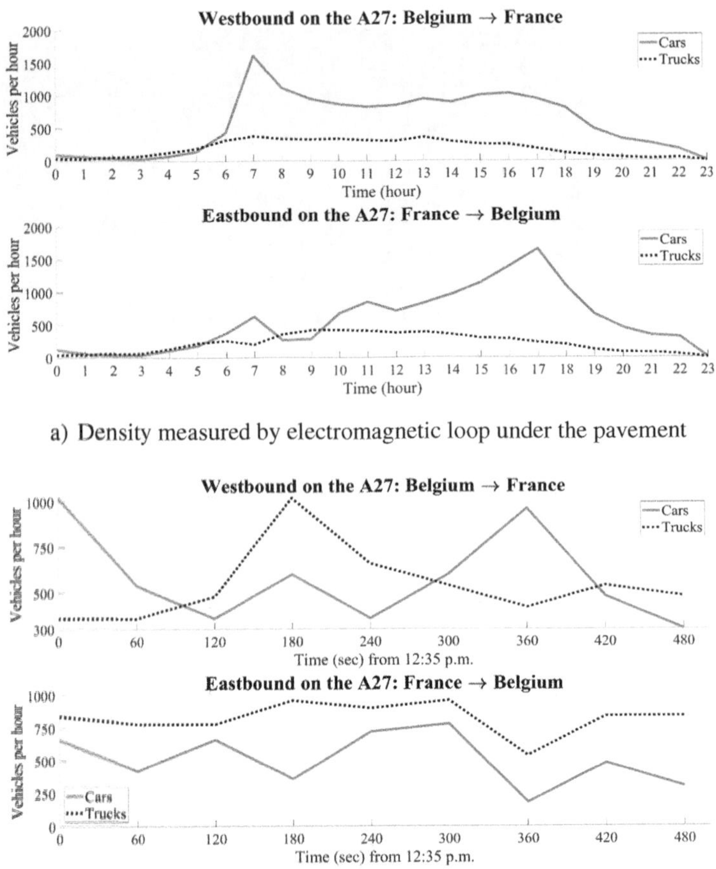

a) Density measured by electromagnetic loop under the pavement

b) Density measured by radar from 12:35 p.m.

Figure 3.14. *Total vehicles per hour on the A27 on April 6, 2017: (a) between midnight and 11 p.m.; (b) from 12:35 p.m. over a period of 8 min (data source: DIR Nord, France) (Rivoirard 2018). For a color version of this figure, see www.iste.co.uk/wahl/design.zip*

The demand for vehicles on the R4 and R5 networks was specified on the basis of traffic data on the A27 provided by the French Interdepartmental Directorate of Northern Roads (DIR Nord) for Thursday April 6, 2017, a weekday outside the school holiday period (Figure 3.14). The data in Figure 3.14(a) are derived from electromagnetic loops placed under the pavement and correspond to measurements of vehicle flow, aggregated hour by hour, between midnight and 11 p.m. The data in Figure 3.14b show density measurements, recorded by a radar sensor, aggregated every minute (60 s) for the same day of April 6, but starting at 12:35 p.m.

Four scenarios (S10 to S13) were defined from these data, the specifications of which have been reported in Table 3.12. The first two scenarios, of varying densities, were constructed for network R4 (scenario S10) comprising the A27 and RD90 and for network R5 (scenario S11), including only the A27. A constant number of 250 cars per hour was defined per traffic direction on the RD90 for scenario S10. The definition of traffic demand for each direction of traffic on the A27 in scenarios S10 and S11 was specified from variable density data measured from 12:35 p.m. and presented in Figure 3.14(b).

Scenario	Network		A27 density (veh/h)				RD90 density (veh/h)			
			O-E direction		E-O direction		O-E direction		E-O direction	
	A27	RD90	Car	Truck	Car	Truck	Car	Truck	Car	Truck
S10	R4 ✓	✓	Variable density				250	0	250	0
S11	R5 ✓		Variable density				0	0	0	0
S12	R5 ✓		630	208	1,620	380	0	0	0	0
S13	R5 ✓		380	260	430	320	0	0	0	0

Table 3.12. *Scenarios S10 to S13 defined on a section of the A27 highway according to different road traffic densities (Rivoirard 2018)*

Scenarios S12 and S13 were defined from the data in Figure 3.14(a). By observation of the measurement curves in this figure, high variability of traffic is visible depending on the time of day and the direction of traffic. For scenarios S12 and S13, fixed vehicle densities were adopted. That of the S12 scenario corresponds to the density of vehicles observed on the A27 highway at peak traffic time of 7 a.m., and the one relatively low of scenario S13 corresponds to the traffic demand observed in the morning of 6 a.m. (Table 3.12).

The construction of mobility traces for each of the nine scenarios S1 to S9 of the R1 to R3 highway networks (Table 3.11) and the four scenarios S10 to S13 of the R4 and R5 networks (Table 3.12) must now be finalized with the SUMO simulator (Figure 3.13) taking into account the traffic demand and the target road network. For this purpose, the Krauß car-following model (Krauß 1998) proposed in SUMO was used. Modeling a vehicle following behavior with the Krauß car-following model would be close to that which would exhibit a driver in response to the actions of the vehicle ahead (Costeseque 2011): a leading vehicle compels the following vehicle to adopt a lower speed than the one desired. Therefore, in this Krauß model, the movements of each vehicle are described according to the positions and speeds of the vehicles in the vicinity and each vehicle tries to reach its "desired speed" while maintaining a perfect safety distance from the vehicle ahead. In SUMO, it is necessary to model the "desired speeds" (*maxSpeed*) of vehicles by speed distributions (Figure 3.15) in order to prevent all vehicles from having the same speed at a given time, and that no overtaking of vehicles be then possible. To this end, the *speedFactor* and *speedDev* parameters in SUMO were specified. Furthermore, the definition for light-duty vehicles of $speedFactor = 1$ and $speedDev = 0.1$ results in a speed distribution in which 95% of these vehicles drive at a speed ranging from 80% to 120% of the legal speed limit (Figure 3.15).

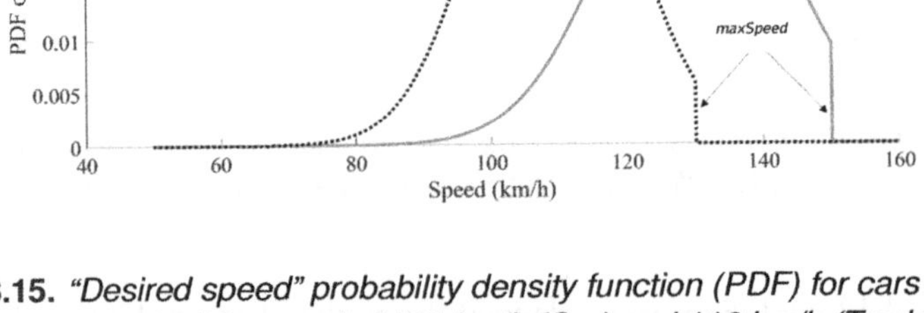

Figure 3.15. *"Desired speed" probability density function (PDF) for cars and trucks for a maximal permissible speed of 130 km/h (Car) and 110 km/h (Truck) (Rivoirard 2018). For a color version of this figure, see www.iste.co.uk/wahl/design.zip*

The full set of the Krauß parameter values for cars and trucks are summarized in Table 3.13. These parameters are *accel*, the acceleration capacity of vehicles; *decel*, their deceleration capacity; *length*, vehicle length; *sigma*, driver imperfection (between 0 and 1); *tau*, the driver's desired intervehicular time; *minGap*, the minimal intervehicular space in case of congestion; *maxSpeed*, the maximal speed of vehicles; *speedFactor*, the multiplier factor of the maximal speed that enables exceeding the speed limit; and finally, *speeDev*, the standard deviation of the *speedFactor* parameter.

Parameters	*accel*	*decel*	*lenght*	*sigma*	*tau*	*minGap*	*maxSpeed*	*speedFactor*	*speedDev*
Units	m/s^2	m/s^2	m	–	s	m	km/h	–	–
Cars	2	3	5	0.5	1	2.5	150	1	0.1
Trucks	1	2	15	0.5	1	5	130	0.84	0.1

Table 3.13. *Krauß car-following model parameters for cars and trucks*

```
<routes>
  <vType id="Car" accel="2.0" decel="3.0" length="5.0" minGap="2.5" maxSpeed="36.1" sigma="0.5"
   speedFactor="1" speedDev="0.1"/>
  <vType id="Truck" accel="1.0" decel="2.0" length="15.0" minGap="5" maxSpeed="30" sigma="0.5"
   speedFactor="0.84" speedDev="0.1" guiShape="truck" />

  <flow id="type1" begin="1" end="3600" vehsPerHour="500" type="Car" departSpeed="max" departLane="random">
       <route edges="1to2"/>
       </flow>
  <flow id="type2" begin="1" end="3600" vehsPerHour="500" type="Car" departSpeed="max" departLane="random">
       <route edges="2to1"/>
       </flow>
  <flow id="type3" begin="1" end="3600" vehsPerHour="100" type="Truck" departSpeed="max" departLane="0">
       <route edges="1to2"/>
       </flow>
  <flow id="type4" begin="1" end="3600" vehsPerHour="100" type="Truck" departSpeed="max" departLane="0">
       <route edges="2to1"/>
       </flow>
</routes>
```

Figure 3.16. *Script for the definition of vehicle demand for the low-density scenario S4 on the R2 network (.rou.xml): four vehicle flows are defined, type1 (Car) and type3 (Truck) flows for one of the traffic directions, type2 (Car) and type4 (Truck) flows for the opposite direction*

An example of a "rout.xml" script file containing the road traffic request and specifications for the Krauß car-following model is shown in Figure 3.16. It corresponds to the vehicle demand of low-density scenarios S1, S4 and S7 for the R1 to R3 networks. Once this file has been created, the SUMO simulation can be run to obtain the vehicle traces with on input the vehicle and road network application files generated, according to the diagram in Figure 3.13. Figure 3.8 shows an illustration, in black on a white background, of the R4 network generated in SUMO with the defined road traffic in yellow.

3.5.1. *Mobility modeling with OPNET* Riverbed Modeler

The traces generated with SUMO for each of the mobility scenarios S1 to S13 cannot be directly imported by the OPNET network simulator *Riverbed Modeler*. In addition, each of the mobility models created contains many vehicle traces (Tables 3.11 and 3.12): for example, 2,400 for S2 and 4,800 for S5 and S8, the three medium-density scenarios, or 5,200 for the high-density scenarios S6 and S9. Three issues can be identified:

1) The format of the SUMO simulator output traces is not compatible with the OPNET trajectory format. These traces therefore need to be ported.

2) The number of mobile nodes simultaneously possible in the simulation with OPNET is limited by the memory capacities of the host machine that hosts our version of OPNET *Riverbed Modeler*. A device should be implemented in order to be able to execute the envisaged scenarios.

3) Finally, the large number of vehicle traces does not allow each vehicle to be manually assigned to a mobile node in the vehicle network. A method for downloading traces and automatically assigning trajectories to nodes must be implemented.

How these points should be solved is now addressed.

3.5.1.1. *Converting SUMO mobility traces into OPNET trajectory*

The OPNET network simulator provides different ways to create and then assign a trajectory to a node. For example, it can be manually created using the Path Editor in OPNET *Riverbed Modeler*. It can also be predefined by the user in a text file with the ".trj" extension, then manually assigned to a node through its parameterization. The OPNET *Riverbed Modeler* user can also resort to a mobility manager that will automatically assign a random trajectory to each of the mobile nodes of the modeled communication network during the initialization phase that follows the launching of a simulation.

An OPNET *Riverbed Modeler* trajectory includes information about the sequence of segments that any node will travel to, which it was assigned by parameterization at the initialization of the simulation; it corresponds to the trace that this (or these) mobile node(s) will follow during the simulation. For each position traveled by the node, it includes the Cartesian coordinates of this position (x, y, altitude), the traveling time to cover the distance separating the previous position from the current position, the waiting time in the current position, and finally the pitch, yaw and roll information that the moving node may undergo. Moreover, it can be parameterized so that to be executed several times during the simulation, and also to be followed in the opposite direction after a potential delay, as shown in Figure 3.17, by the contents of the file of a trajectory in the ".trj" format defined from the first three positions (*Coordinate_Count* to 3) of the "type1.0" vehicle of the ".xml" file of the SUMO mobility traces (Figure 3.18).

```
Version: 5
Position_Unit: Meters
Altitude_Unit: Meters
Coordinate_Method: fixed
Altitude_Method: absolute
locale: C
Calendar_Start: unused
Execute_Count: 1
Reverse_Trajectory: no
Reverse_Wait_Time: 0.000000
Coordinate_Count: 3
# X Position ,Y Position ,Altitude ,Traverse Time ,Wait Time ,Pitch ,Yaw ,Roll
5.100000 ,-1.650000 ,0 ,0h0m0.1s ,0h0m0.0s ,Autocomputed ,Autocomputed ,Unspecified
40.610000 ,-1.650000 ,0 ,0h0m1s ,0h0m0.0s ,Autocomputed ,Autocomputed ,Unspecified
76.160000 ,-1.650000 ,0 ,0h0m1s ,0h0m0.0s ,Autocomputed ,Autocomputed ,Unspecified
```

Figure 3.17. *Example of a ".trj" trajectory in the OPNET Riverbed Modeler format. For a color version of this figure, see www.iste.co.uk/wahl/design.zip*

```
<timestep time="0.00"/>
<timestep time="1.00">
  <vehicle id="type1.0" x="5.10" y="-1.65" angle="90.00" type="Car" speed="36.10" pos="5.10" lane="1to2_2" slope="0.00"/>
  <vehicle id="type2.0" x="4994.90" y="4.95" angle="270.00" type="Car" speed="33.87" pos="5.10" lane="2to1_1" slope="0.00"/>
  <vehicle id="type3.0" x="15.10" y="-8.25" angle="90.00" type="Truck" speed="30.00" pos="15.10" lane="1to2_0" slope="0.00"/>
  <vehicle id="type4.0" x="4984.90" y="8.25" angle="270.00" type="Truck" speed="30.00" pos="15.10" lane="2to1_0" slope="0.00"/>
</timestep>
<timestep time="2.00">
  <vehicle id="type1.0" x="40.61" y="-1.65" angle="90.00" type="Car" speed="35.51" pos="40.61" lane="1to2_2" slope="0.00"/>
  <vehicle id="type2.0" x="4961.07" y="4.95" angle="270.00" type="Car" speed="33.83" pos="38.93" lane="2to1_1" slope="0.00"/>
  <vehicle id="type3.0" x="45.08" y="-8.25" angle="90.00" type="Truck" speed="29.98" pos="45.08" lane="1to2_0" slope="0.00"/>
  <vehicle id="type4.0" x="4955.15" y="8.25" angle="270.00" type="Truck" speed="29.75" pos="44.85" lane="2to1_0" slope="0.00"/>
</timestep>
<timestep time=
```

Figure 3.18. *Excerpt from the beginning of the ".xml" trace file obtained after simulation with SUMO for the low-density S4 scenario on R2. The time step separating two timestep times consecutively is 1 s. For a color version of this figure, see www.iste.co.uk/wahl/design.zip*

A SUMO simulation generates a list of mobility traces at each time step of the set of vehicles on the road network. This list is saved in an ".xml" file in the *Floating*

Car Data (FCD) format. This file thus contains for each date (i.e. for each sampling step) the information relating to each vehicle existing in the road network, namely, *vehicle id*, the unique identification, the Cartesian coordinates in x and y, the speed, the vehicle angle and the type of vehicle (*truck* or *car* in the simulated scenarios), the position, lane and slope gradient. Figure 3.18 shows the excerpt of the data obtained at times 1 and 2 s for the low-density S4 scenario on the R2 network.

To use SUMO-defined vehicle traces within *Riverbed Modeler*, the trajectory of each vehicle on the road network must be extracted from the ".xml" file to be converted to the *Riverbed Modeler* format and saved in a file with a ".trj" extension (Figure 3.17). We carried out this operation in two stages.

First, the ".xml" trace file is imported into Excel (Figure 3.19) with a replacement of the names of vehicle identifiers (type1.0, type2.0, etc.) in number. In this example, only the date, vehicle identification and position values (x, y) are useful for constructing the trajectories in OPNET format, since the road gradient was not considered when modeling the road network. These values are stored in a tabulate file in text format, which can be downloaded by any type of programming language.

	A	B	C	D	E	F	G	H	I	J
1	time	id	x	y	angle	type	speed	pos	lane	slope
2	1	100000	5.1	-1.65	90	Car	36.1	5.1	1to2_2	0
3	1	200000	4994.9	4.95	270	Car	33.87	5.1	2to1_1	0
4	1	300000	15.1	-8.25	90	Truck	30	15.1	1to2_0	0
5	1	400000	4984.9	8.25	270	Truck	30	15.1	2to1_0	0
6	2	100000	40.61	-1.65	90	Car	35.51	40.61	1to2_2	0
7	2	200000	4961.07	4.95	270	Car	33.83	38.93	2to1_1	0
8	2	300000	45.08	-8.25	90	Truck	29.98	45.08	1to2_0	0
9	2	400000	4955.15	8.25	270	Truck	29.75	44.85	2to1_0	0
10	3	100000	76.16	-1.65	90	Car	35.55	76.16	1to2_2	0

Figure 3.19. *Excerpt of the resulting import into an Excel spreadsheet of the trace file obtained with SUMO. Identifiers in the "typek.l" format (k and l are integers) have been converted into a number. For a color version of this figure, see www.iste.co.uk/wahl/design.zip*

In a second step, the *Riverbed Modeler* trajectories are generated by a routine that extracts for each vehicle the set of Cartesian coordinates (x,y) of its trace, counts the number of positions contained in this trace (i.e. the value of $Coordinate\_Cout$), then writes the header and reports line by line the information relating to the successive coordinates obtained. The *Traverse_Time* value is 1 s (which corresponds to the time interval between two successive vehicle coordinates), except for the first coordinate (section 3.5.1.2). The $Wait\_Time$ time allows the trajectory to be initiated on the start date of the SUMO trace (vehicle entry date in road network). At the end of this second time, a trajectory file in the ".trj" format was generated per vehicle present in the ".xml" file.

3.5.1.2. *Workaround for the node limit in OPNET*

The previously identified vehicle requirements (Tables 3.11 and 3.12) are greater than the number of communication nodes that can be initialized when running the OPNET *Riverbed Modeler* simulation. Compute and memory capacities are actually limited by the host machine hosting the OPNET license. As such, Rivoirard (2018) experienced a major crash with the simulator software when modeling 666 nodes on average for the purposes of supporting a simulation of scenario S5 on R2 during 500 ms for a total medium density of 4,800 vehicles per hour in the network (Table 3.11). A solution was adopted to, on the one hand, meet the requirements set in terms of vehicle density in the road network, and on the other, to reduce the number of nodes to be introduced into the communication network in order to model all these vehicles during the simulation. The underlying idea is that the same communication node should be successively used for different vehicles during the simulation. This solution is based on the definition of a buffer zone, geographically distant from the area modeling the road network space, and the concatenation of several different vehicle traces within the same OPNET trajectory file (section 3.5.1.2.1). It also requires adapting the state model of the network protocol in order to reset its own variables before a new trace is executed by a mobile node (section 3.5.1.2.2).

3.5.1.2.1. Building multi-trace trajectory files

In order to reduce the number of mobile nodes to be modeled while keeping the same density scenarios previously defined, several different vehicle traces must be concatenated into a single OPNET trajectory file. Between two traces, the mobile node is placed inside a buffer zone, also known as the garage area. *This buffer zone is critical to the quality of the simulation results:* it keeps the moving nodes out of communication range of the road area between the execution of two separate vehicle traces so that the metric statistics, measured on the 0–5 km portion of the trajectory, are not impacted by the transmission of messages from the nodes then waiting for a change of function. For networks R1 to R3, defined longitudinally between the 0 and 5 km abscissas, the buffer zone is specified by two points. A first point is positioned at Cartesian coordinates (10,000, 10,000), expressed here in meters, and corresponds to the exit point of the mobile node to reach the beginning of the road section, located either at the 0 or 5 km abscissa, depending on the direction of the trajectory. The second point is positioned at coordinates expressed in meters (10,000, -10,000) and corresponds to the entry point to the buffer zone that the mobile node reaches when it leaves the highway section after reaching the x-axis 5 km, respectively 0 km, as illustrated in white by the multi-trace trajectory shown in Figure 3.20.

The purpose of concatenating multiple vehicle traces into a single ".trj" trajectory file is to allow the use of the same mobile node for several vehicles during simulation. The modeled node will travel one by one to all of the traces contained in the trajectory file assigned to it at initialization. At the end of a trace, the node is brought back to the location of the buffer zone, where it is placed in a queue until the next trace start date. Figure 3.21 illustrates the format of a ".trj" trajectory obtained by the serialization of a set of vehicle traces existing on the road network during strictly disjunct periods of time. Algorithm 3.1 describes the routine implemented for specifying such multi-trace trajectories resulting from processing the SUMO data obtained in text format after importing into Excel spreadsheet (Figure 3.19). Section 3.5.1.2.2 specifies when the neighbor and routing tables are reset before a new trace is reassigned to the node.

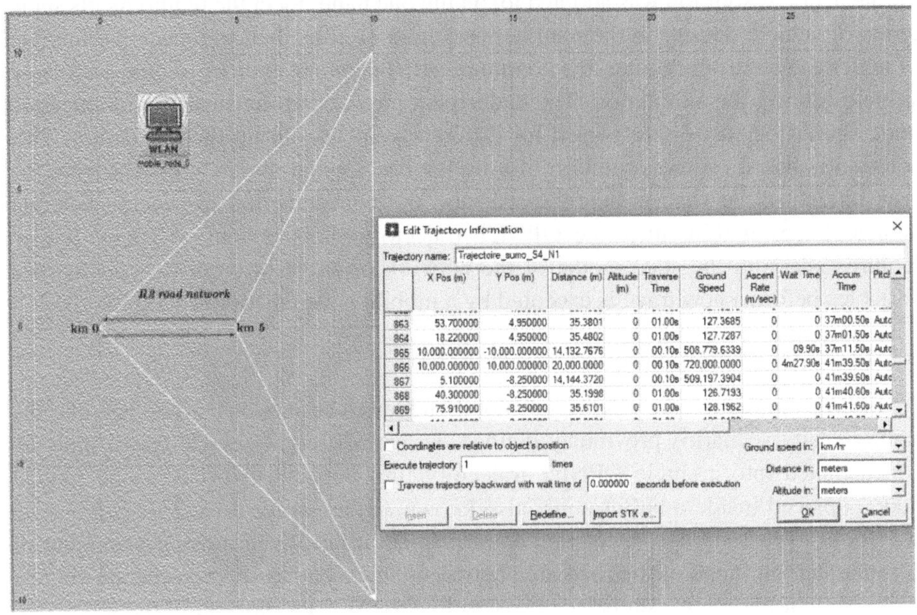

Figure 3.20. *Multi-trace trajectory of a mobile node on the R2 road network edited in OPNET Riverbed Modeler. For a color version of this figure, see www.iste.co.uk/wahl/design.zip*

The ".trj" trajectories obtained with Algorithm 3.1 contain up to a dozen traces of different vehicles. As a result, for example, only 62 nodes are needed for modeling a density of 600 vehicles per hour for scenario S1 on the R1 network or 532 nodes for a density of 4,800 vehicles per hour for scenario S5 on the R2 network (Table 3.14). Despite this strategy, the simulation of S6 and S9 scenarios was not possible due to the hardware configuration of the host machine, which would require, respectively, modeling 999 and 1,454 nodes.

```
Version: 5
Position_Unit: Meters
Altitude_Unit: Meters
Coordinate_Method: fixed
Altitude_Method: absolute
locale: C
Calendar_Start: unused
Execute_Count: 1
Reverse_Trajectory: no
Reverse_Wait_Time: 0.000000
Coordinate_Count: 1297
# X Position ,Y Position ,Altitude ,Traverse Time ,Wait Time ,Pitch ,Yaw ,Roll
10000 ,10000 ,0 ,0h0m0.1s ,0.900000 ,Autocomputed ,Autocomputed ,Unspecified
5.100000 ,-1.650000 ,0 ,0h0m0.1s ,0h0m0.0s ,Autocomputed ,Autocomputed ,Unspecified
40.610000 ,-1.650000 ,0 ,0h0m1s ,0h0m0.0s ,Autocomputed ,Autocomputed ,Unspecified
76.160000 ,-1.650000 ,0 ,0h0m1s ,0h0m0.0s ,Autocomputed ,Autocomputed ,Unspecified
[...]
4955.540000 ,-8.250000 ,0 ,0h0m1s ,0h0m0.0s ,Autocomputed ,Autocomputed ,Unspecified
4991.200000 ,-8.250000 ,0 ,0h0m1s ,0h0m0.0s ,Autocomputed ,Autocomputed ,Unspecified
10000 ,-10000 ,0 ,0h0m0.1s ,0h0m9.9s ,Autocomputed ,Autocomputed ,Unspecified
10000 ,10000 ,0 ,0h0m0.1s ,259.900000 ,Autocomputed ,Autocomputed ,Unspecified
4994.900000 ,8.250000 ,0 ,0h0m0.1s ,0h0m0.0s ,Autocomputed ,Autocomputed ,Unspecified
4960.880000 ,8.250000 ,0 ,0h0m1s ,0h0m0.0s ,Autocomputed ,Autocomputed ,Unspecified
4926.770000 ,8.250000 ,0 ,0h0m1s ,0h0m0.0s ,Autocomputed ,Autocomputed ,Unspecified
[...]
49.130000 ,8.250000 ,0 ,0h0m1s ,0h0m0.0s ,Autocomputed ,Autocomputed ,Unspecified
15.280000 ,8.250000 ,0 ,0h0m1s ,0h0m0.0s ,Autocomputed ,Autocomputed ,Unspecified
10000 ,-10000 ,0 ,0h0m0.1s ,0h0m9.9s ,Autocomputed ,Autocomputed ,Unspecified

[deleted lines,
then first and last lines of last stacked trajectory...]

10000 ,10000 ,0 ,0h0m0.1s ,263.900000 ,Autocomputed ,Autocomputed ,Unspecified
5.100000 ,-1.650000 ,0 ,0h0m0.1s ,0h0m0.0s ,Autocomputed ,Autocomputed ,Unspecified
41.060000 ,-1.650000 ,0 ,0h0m1s ,0h0m0.0s ,Autocomputed ,Autocomputed ,Unspecified
[...]
4946.640000 ,-8.250000 ,0 ,0h0m1s ,0h0m0.0s ,Autocomputed ,Autocomputed ,Unspecified
4982.200000 ,-8.250000 ,0 ,0h0m1s ,0h0m0.0s ,Autocomputed ,Autocomputed ,Unspecified
10000 ,-10000 ,0 ,0h0m0.1s ,0h0m9.9s ,Autocomputed ,Autocomputed ,Unspecified
```

Figure 3.21. *".trj" file contents of the trajectory of a node consisting of a stack of traces of various vehicles. For a color version of this figure, see www.iste.co.uk/wahl/design.zip*

Density	Low			Medium			High			Other			
Network	R1	R2	R3	R1	R2	R3	R1	R2	R3	R4		R5 (A27)	
Scenario	S1	S4	S7	S2	S5	S8	S3	S6	S9	S10	S11	S12	S13
Modeled nodes	62	138	123	262	532	522	567	*999*	*1,454*	380	328	357	175

Table 3.14. *Number of nodes modeled per scenario in Riverbed Modeler*

Require: A $Traces.txt$ file in tabulated text format containing all the mobility traces extracted from the Sumo trace file to be converted to .trj format (Figure 3.19)
Ensure: .trj trajectory files

//**Initializations**
$TimeStep \leftarrow 1\text{s}$ ▷ time step between two successive positions in the same trace
//Definition of a geographical area outside the communication range of the road section, referred to as the garage area (P, *Parking*)
$X\_P \leftarrow 10000$
$Y\_InP \leftarrow 10000$
$Y\_OutP \leftarrow -10000$
$Alt \leftarrow 0$ ▷ *Alt, Altitude*
//Constants relating to node mobility between two positions
$TT\_InOutP \leftarrow 0h0m0.1s$ ▷ Traverse time towards or out of the garage area to the first position of the new trace (TT, *Traverse Time*)
$TT\_NextPos \leftarrow 0h0m1s$ ▷ Traverse time to the next trace position (TT, *Traverse Time*) for a $TimeStep$ of 1s
$WT\_AtPos \leftarrow 0h0m0s$ ▷ Waiting time at the current position of a vehicle trace before moving on to the next one (WT, *Wait Time*)
$C\_Parking \leftarrow 20s$ ▷ Cumulative wait time in the garage area
$WT\_P \leftarrow \frac{1}{2}C\_Parking - 0.1s$ ▷ Wait time at position (X_P, Y_OutP)
$Pt \leftarrow Autocomputed$ ▷ *Pt, Pitch*
$Yaw \leftarrow Autocomputed$ ▷ Yw, *Yaw*
$Rl \leftarrow Unspecified$ ▷ *Rl, Roll*
//**Begin**
READ "*trace.txt*" ▷ structured tabulated text file containing the date, identification, X and Y positions information of the traces extracted from the Sumo trace file (Figure 3.19)
$\mathcal{ID} \leftarrow Im(\mathcal{ID})$ ▷ $\mathcal{ID}$, the set of vehicle identification numbers, contained in $Traces.txt$, sorted by date of entry into the road network
$NVEH \leftarrow Card(\mathcal{ID})$ ▷ $NVEH$, number of different traces (that is, vehicles) contained in the $Traces.txt$ file
// Search for the number of trajectory files to be created
$Max\_veh\_on\_Road \leftarrow 0$ ▷ Initialization of the number of mobile nodes and therefore of trajectory files necessary
for $veh <= NVEH$ with a step of 1 **do**
    $StartT \leftarrow ID(veh)$ trace start date
    $EndT \leftarrow ID(veh)$ trace end date
    $ID\_veh\_on\_Road \leftarrow$ number of separate traces present in the interval $[StartT, EndT + C\_Parking]$

```
    if Max_veh_on_Road < ID_veh_on_Road then
        Max_veh_on_Road ← ID_veh_on_Road
    end if
end for
// Computation of the number of positions per .trj trajectory
for i <= Max_veh_on_Road with a step of 1 do
    nb ← 0
    Coordinate_Count ← 0
    for veh <= NVEH with a step of Max_veh_on_Road do
        tmp ← Trace entry number ID(veh)
        nb ← nb + tmp + 2    ▷ Two wait positions, at the start and end of trace
    end for
    Coordinate_Count ← nb    ▷ Number of positions in the i-th trajectory file
end for
//Trajectories files creation
for i <= Max_veh_on_Road with a step of 1 do    ▷ Creation of the i-th .trj trajectory file
    for veh <= NVEH with a step of Max_veh_on_Road do
        Data ← traces of vehicle ID(veh)
        if veh = i then
            DateFinTrace ← 0
            Create ASCII file name: Fichier_Traj_i.trj
            OPEN: Fichier_Traj_i.trj
            WRITE file header and number of positions Coordinate_Count    ▷ see header of Figure 3.17 or 3.21
        end if
        DateDebTrace ← trace start date
        Wait_Time ← DateDebTrace − DateFinTrace − 0.1
        WRITE first node position: X_P , Y_InP , Alt , TT_InOutP , Wait_Time , Pt , Yw , Rl
        for each line of Data do    ▷ Write the following positions
            if First line of Data then
                Traverse_Time ← TT_InOutP
            else
                Traverse_Time ← TT_NextPos
            end if
            Wait_Time ← WT_AtPos
            X ← XCoordinate
            Y ← YCoordinate
            WRITE: X , Y , Alt , Traverse_Time , Wait_Time , Pt , Yw , Rl
        end for
```

```
        WRITE: X_P , Y_OutP , Alt , TT_InOutP , WT_P , Pt , Yw , Rl
    end for
    CLOSE: Fichier_Traj_i.trj
end for
```

Algorithm 3.1. *Creation of multi-trace trajectory files with .trj format*

3.5.1.2.2. Adaptation of the *Process Model* of routing protocols to multi-trace trajectories

Each modeled communication node now executes a trajectory that successively describes the single paths of separate vehicles, which arrive and then cross the road section of interest during separate time periods. On the other hand, during the simulation, the network simulator collects local statistics for each node. In order for these statistics to correspond to the scenarios played on the road section, it is necessary to stop measurements when the node is located outside that road section. Furthermore, the node routing tables must be flushed of any information collected during the past trace.

The OLSR, OLSR-CBL and AODV routing protocols evaluated in Chapter 4 are each available as a *Process Model* represented by states and transitions. For example, Figure 3.22(a), the first state of the OLSR *Process Model*, namely the "init" state, performs the initialization of the protocol and associated interfaces. The process then transitions to the "wait" state, where it reacts to events such as when receiving HELLO and topology control (TC) messages or timeout messages associated with the computation of routes (Sondi 2010).

The example of the OLSR *Process Model* (Figure 3.22(b)) shows that a "initialization" state has been added between the "wait" state and the "init" state. The C code of the process states has been modified accordingly. The transition from the "initialization" state to the "init" state is done by performing an "insert" condition and the transition from the "wait" state to the "initialization" state when achieving an "reset" condition. The input state of the modified OLSR process model when the simulation is started is now the "initialization" state. In this state, the process retrieves the dimensions of the (road) network of interest (Xmin, Xmax, Ymin, Ymax) and then checks whether the node position is within that area. If this is the case, the process exits the state and enters the "init" state, where the shared global variables, state variables and statistics of OLSR are reset. The process then continues in the "wait" state, where it executes events related to the routing protocol as long as the node does not leave the road area of interest. When its exit from this zone is detected, a routing protocol parameter reset function is executed. This cancels pending events related to the timers of HELLO and TC messages, clears the contents of the various routing tables (routing, neighbor, MPR tables, etc.), updates the statistic related to the node status if necessary (the node loses its MPR status if it had it) and then the process enters the "initialization" state, from which it will only exit when the node is again positioned in the geographic area of interest.

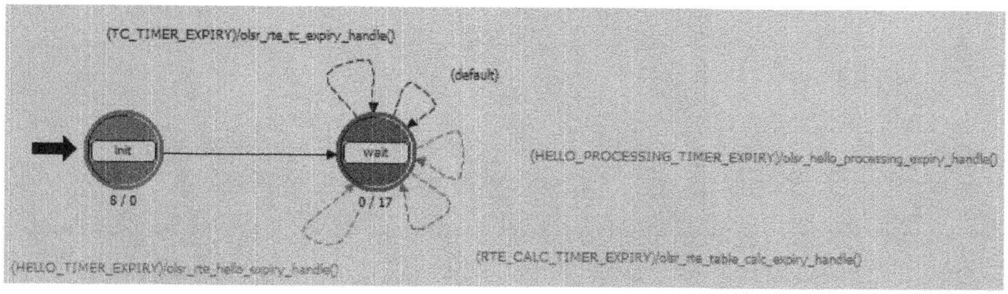

a) OLSR *Process Model* in OPNET *Riverbed Modeler*

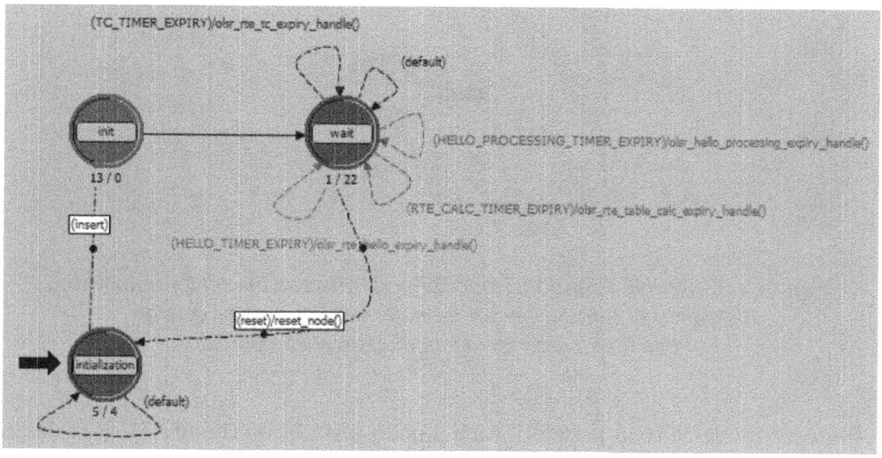

b) Modified OLSR *Process Model*

Figure 3.22. *OLSR Process Model before and after adding a function for inserting and resetting a mobile node during the simulation. For a color version of this figure, see www.iste.co.uk/wahl/design.zip*

3.5.1.3. *Trajectory automatic download method in OPNET*

The creation of multi-trace trajectories results in reducing the number of nodes in simulations (Table 3.14). However, there still remain a large number of modeled nodes to which a trajectory should be assigned. One way to proceed is explained hereafter.

For a given scenario (S1 to S13), the set of all the network nodes in the scenario are placed in a simulation space; in this case, a campus with a size of

10 km × 10 km. Figure 3.20 shows a node positioned in such a campus and the trajectory that was manually assigned to it. When a scenario includes hundreds of nodes, manually assigning a trajectory to each of them makes little sense. OPNET, however, provides a *mobility_node* node model (OPNET *Mobility Config* node model, Figure 3.23) whose *random_mobility_mgr* process makes it possible to automatically allocate a random vector path to each node located inside the campus.

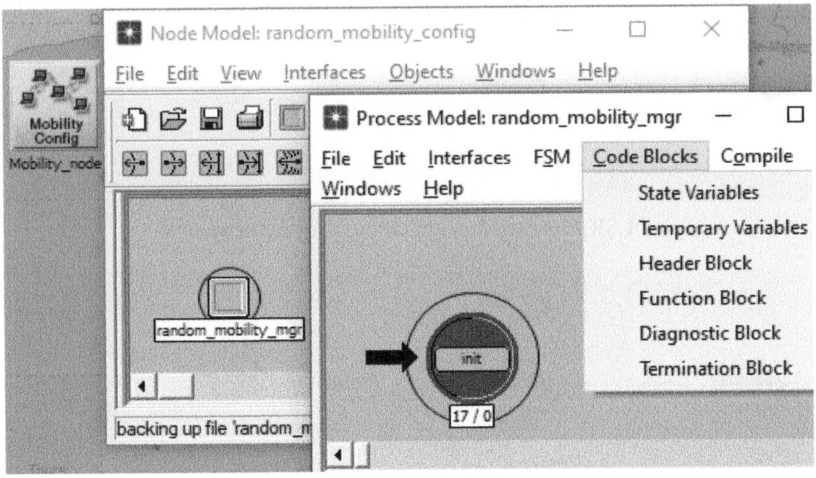

Figure 3.23. *OPNET random_mobility_mgr process model for automatic allocation of random vector trajectories. For a color version of this figure, see www.iste.co.uk/wahl/design.zip*

For each modeled road scenario, such a node was placed (Figure 3.25) at the top hierarchical level of the location where the *Campus_Network* subnet is located (subnet where the set of all the mobile nodes necessary for the execution of the road scenario being considered are placed in bulk). The code for the process model of this node was modified with the addition of a "set_position_of_nodes()" function (Figure 3.24), whose role is to download previously created trajectories from a working directory and automatically assign them to the *Campus_Network* nodes. In addition, two attributes were added to the mobility node renamed "Trajectory_mobility_node_0" (Figure 3.25): the "Trajectory Directory" attribute is used to specify the path to the directory where the created trajectories are located, the "File Name Radical" attribute is for defining the radical of the file name from which the "set_position_of_nodes()" function derives the set of trajectory file names. These two attributes were defined at the node process model level, but parameterized as "Promote" in order to be configurable at the higher hierarchical level, namely at the attribute level of the mobility node. Algorithm 3.2 shows the C code of the "set_position_of_nodes()" function. This function retrieves the name of the trajectory files and the path to access them, and obtains the number num_nodes of mobile nodes modeled at the subnet level; finally, for any node i in the subnet numbered from 0 to $num\_nodes - 1$, it assigns it the trajectory number $i + 1$ and positions it at the first coordinate defined in that trajectory file.

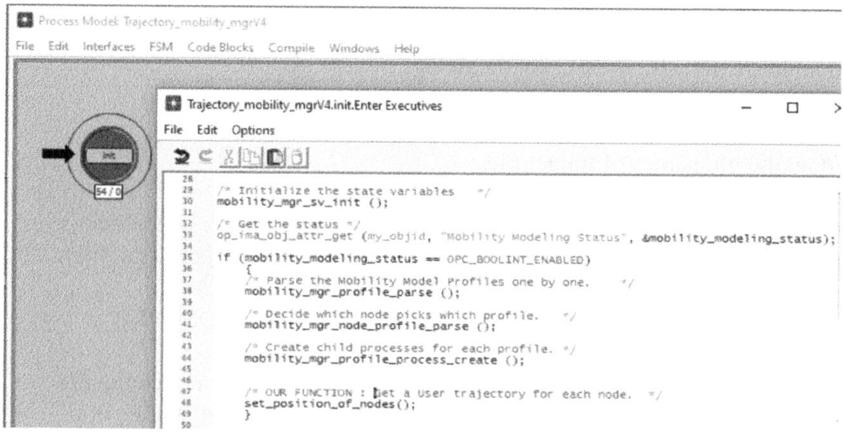

Figure 3.24. *Trajectory_mobility_mgr.pr.m, a modified release of the OPNET random_mobility_mgr Process Model to assign a trajectory to each of the mobile nodes in the subnet. For a color version of this figure, see www.iste.co.uk/wahl/design.zip*

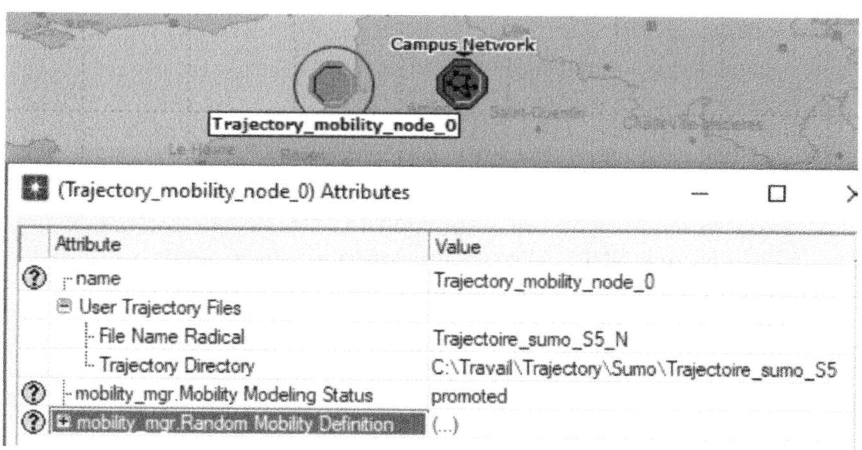

Figure 3.25. *Trajectory_mobility_node Process Model attributes in OPNET. For a color version of this figure, see www.iste.co.uk/wahl/design.zip*

```
static void set_position_of_nodes(void) {
Objid node_objid;
int i, inc, cmp, num_nodes;
unsigned int buffer_size=80;
char traj[80], directory[80], separator[3], traj_file[80];
char path[300], path2[300], num_traj[50], ligne[800];
FILE * file;
double src_nx, src_ny, src_alt,src_delay;
char *suiteChaine = NULL; char *value;
FIN (set_position_of_nodes<args>);
//Get the file names of trajectories:
ppid=op_topo_parent(op_id_self());
op_ima_obj_attr_get_str(ppid,"Trajectory Directory",buffer_size,directory);
op_ima_obj_attr_get_str(ppid,"File Name Radical",buffer_size,traj_file);
//Get the Number of mobile nodes in the system:
num_nodes = op_topo_object_count (OPC_OBJTYPE_NDMOB);
for (i = 0; i < num_nodes; i++) {
    //Get node ID, build the name of the i_th trajectory, then open the file
    node_objid = op_topo_object (OPC_OBJTYPE_NODE_MOB, i);
    sprintf (separator,"\\",""); strcpy(path,directory); strcat(path,separator);
    strcpy(traj,traj_file); sprintf(num_traj,"%d",i+1); strcat(traj,num_traj);
    strcat(path,traj); sprintf(path2,"%s.trj",path);
    file = fopen(path2, "r");
    inc=0;
    while (inc <12) {
        inc=inc+1; fgets(ligne,800,file);
        if (inc==7) {
        suiteChaine = strchr(ligne, ': ');
        cmp=strcmp(suiteChaine, "unused");
        if (cmp==0) {src_delay=0.0; } else {src_delay=atof(suiteChaine); } }
    if (fgets(ligne,500,file) !=NULL) { //Get line 13
        value=malloc(30*sizeof(char)); value=strtok(ligne,",");
        src_nx=atof(value); value=strtok(NULL,","); src_ny=atof(value);
        value=strtok(NULL,","); src_alt=atof(value); free(value);
        //Set trajectory i to node i and the mobile node's position attributes.
        op_ima_obj_attr_set_dbl (node_objid, "x position", src_nx);
        op_ima_obj_attr_set_dbl (node_objid, "y position", src_ny);
        op_ima_obj_attr_set_dbl (node_objid, "altitude", src_alt);
        op_ima_traj_set (node_objid, traj, src_delay); }
    fclose(file); }
FOUT;}
```

Algorithm 3.2. *Function for assigning trajectories to nodes and positioning them at the initial trajectory position in OPNET Riverbed Modeler*

3.6. Transmission modeling

3.6.1. *Choice of technology for frame transmission*

The operation of the CBL *clustering* scheme is independent of the physical and data link layers of the radio technology used. The "dynamically stable" structure constructed by the CBL scheme only takes into account the invariant of the radio range of the technology employed. However, a full evaluation of the protocol under near-field conditions can be performed for any underlying technology chosen.

At the time of the study reported in this book, IEEE 802.11p was the only standard designed for VANETs that allowed direct V2V communication. Admittedly, version 13 of the LTE technology promoted by Europe also enabled at that time *device-to-device* (D2D) communications; however, this is limited to a single hop. Furthermore, they were intended for public safety purposes, on the one hand, and under the coordination of a static terminal called eNodeB to obtain access wireless communication. In the commercial version of LTE, only the discovery phase of D2D is allowed and direct communications without involving an eNodeB are not possible. 5G technology aims to be more flexible and eventually make possible V2V communications (Gruet 2018). Since then, version R15 (from 2018) only supported one-hop direct V2V communications (3GPP 2023). It was only in July 2020 with version R16 that direct multi-hop V2V communications became possible. In the meantime, we have opted for IEEE 802.11p technology, capable of ensuring a transmission range up to 1 km. For all simulations carried out under OPNET *Riverbed Modeler*, we used the IEEE 802.11p model available under the OPNET *Riverbed Modeler* wireless license.

The IEEE 802.11p wireless communication protocol is an amendment to the IEEE 802.11 standard for wireless access in vehicular environments, both for V2V communications and vehicle-to-infrastructure or infrastructure-to-vehicle (i.e. between high-speed entities with high relative speeds). The IEEE 802.11 standard is defined on the physical and data link layers of the ISO OSI reference model.

With respect to the physical layer, 802.11p networks operate on the frequency band 5.850–5.925 GHz in North America and 5.855–5.925 GHz in Europe; signal coding on the medium is an *Orthogonal Frequency-Division Multiple access* (OFDM) encoding (Abdelgader and Lenan 2014). The IEEE 802.11p standard divides the 5.850–5.925 GHz band into seven 10 MHz channels with one control channel and six service channels (IEEE 2010). The control channel is reserved for the transmission of network management messages. The other six channels are dedicated to the transmission of application data broadcast on the control channel. This method is based on the 802.11a standard which uses 20 MHz channels. The 802.11p channel frequency band size was halved compared to 802.11a (which also operates on the 5 GHz frequency band) in order to better address reflected signals and restrict the Doppler effect. Consequently, throughput is also half. Up to eight data transfer rates are possible with the OFDM method, depending on the modulation and encoding methods used (3; 4.5; 6; 9; 12; 18; 24 and 27 Mbit/s). Four categories of *Effective Isotropic Radiated Power* (EIRP) are allowed, ranging from 23 to 44.8 dBm for the United States and from 23 to 33 dBm for Europe.

Regarding the data link layer, with direct D2D ad hoc communications, the IEEE 802.11 medium access control protocol (MAC) executes a distributed carrier detection algorithm called *Carrier Sense Multiple Access with Collision Avoidance* (CSMA/CA, Figure 3.26). The aim of the CSMA/CA access protocol is to defer the transmission of a frame by a random number of time intervals (called *Time Slot*) if at the time of the preemption attempt of the communication medium the latter is not free. In this way, the algorithm reduces the probability of collision between messages from neighboring nodes that would otherwise, without these waiting windows, attempt to transmit concurrently as soon as the medium is free again. This results in preventing message collisions, but their occurrence cannot be completely avoided.

Two communication modes are possible at the MAC level: by *broadcast* or to a single destination node from which is expected an acknowledgment of the frame transmitted by the receiver (*unicast*). Among others, the values of the contention window CW_{min} and CW_{max} and the communication medium monitoring time T_{AIFS}, which are indicated in the flow chart of Figure 3.26 and given in Table 3.15, depend on the access category (AC) chosen for transmitting frames (Bilstrup et al. 2008). The 802.11p standard makes a distinction between four of them that provide different access priorities to the media: AC_BK (ETSI 2020) gives a priority called *background* for a background transmission service when traffic is low; AC_BE is a priority called *best-effort* which provides "a best-effort" service for standard applications; AC_VI is a *video* priority for video, and AC_VO is a *voice* priority for voice (ETSI 2020). When communication is carried out in *unicast* mode, an acknowledgment (ACK frame) transmitted by the destination node is expected in return for the successful receipt of the frame. If it does not receive it, the source node will retry its transmission after doubling the value of $CW(AC)$ of its contention window, while a good reception is not validated and the maximal value of the contention window is not reached. The $CW_{max}(AC)$ value of the protocol is therefore used, both to limit the number of possible transmission attempts in *unicast* mode, and also to gradually decrease the collision probability in the event of an overloaded network by increasing the size of the contention window.

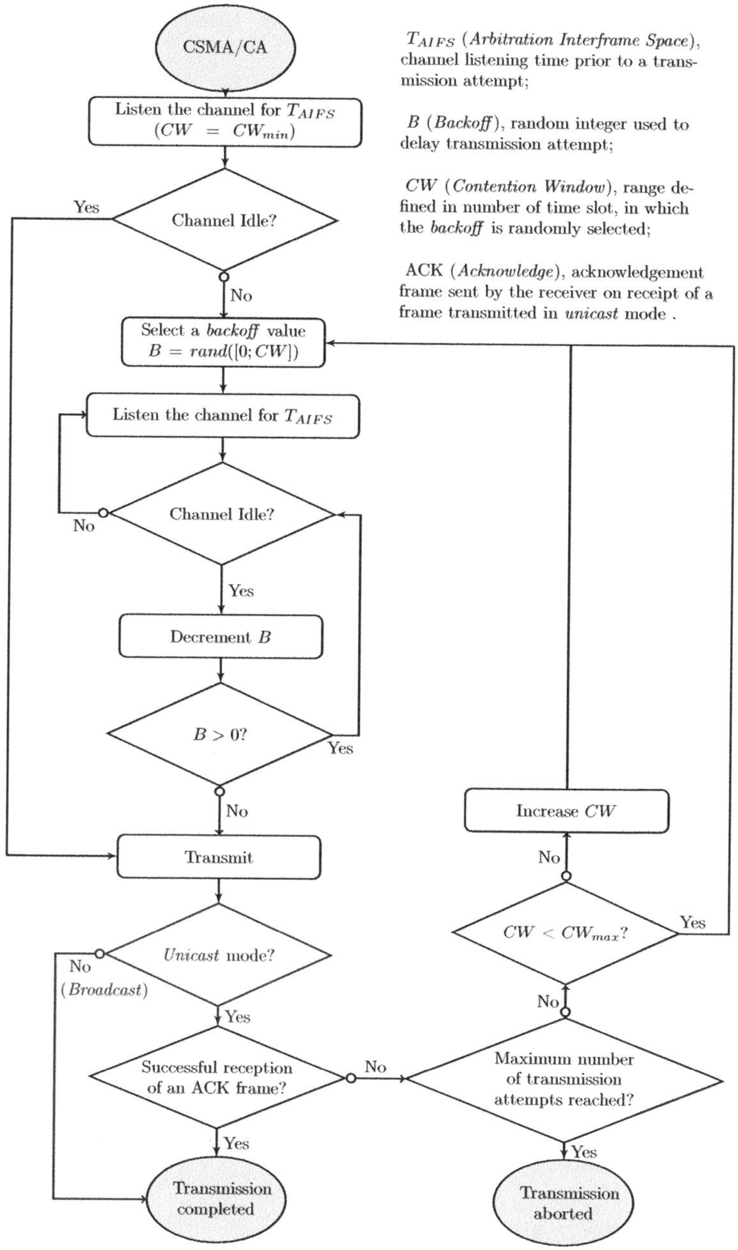

Figure 3.26. *CSMA/CA model – transmission in unicast mode, respectively, broadcast. Diagram based on two (ETSI 2020) flow charts. For a color version of this figure, see www.iste.co.uk/wahl/design.zip*

3.6.2. *Parametrization of an 802.11p node under OPNET* Riverbed Modeler

OPNET *Riverbed Modeler* provides a WLAN station node model (wlan_wkstn_adv) that implements a modeling of the 802.11p radio communication interface. For simulations, the default free-space propagation model was used. The antenna under consideration is omnidirectional and the radio interface of the mobile nodes was configured to achieve a maximum radio communication range of 500 m, which corresponds to the communication range expected by 802.11p radio technology. This relatively short range also makes it possible to confine the required length of motorway mobility scenarios for evaluating CBL to 5 km sections. To achieve this range, the transmission power was set to 0.005 W and the reception threshold to - 95 dBm (Table 3.16). The transmission rate was also set at 12 Mbits/s in line with the operating capabilities of 802.11p WIFI technology. In terms of transmission channel, only the first 10 Mhz channel of the frequency band (5.855–5.925 GHz), allocated in Europe by ETSI, was used for our simulations. In terms of 802.11p access to the channel, two access categories were configured: the *background* access category (AC_BK) assigned to the background transmission of routing messages and *best-effort* (AC_BE) dedicated to the transmission of application messages with a medium access priority slightly higher than that given to routing traffic (Table 3.15).

Category	$CW_{min}(AC)$	$CW_{max}(AC)$	AIFSN(AC)	$T_{AIFS}(AC)$
AC_BK	$aCW_{min} = 15$	$aCW_{max} = 1023$	9	149 μs
AC_BE	$aCW_{min} = 15$	$aCW_{max} = 1023$	6	110 μs
AC_VI	$\frac{1}{2}(aCW_{min} + 1) - 1 = 7$	$aCW_{min} = 15$	3	71 μs
AC_VO	$\frac{1}{4}(aCW_{min} + 1) - 1 = 3$	$\frac{1}{2}(aCW_{min} + 1) - 1 = 7$	2	58 μs
	where $T_{AIFS}(AC) = SIFS + AIFSN(AC) * SlotTime$ with by default $SIFS = 32$ μs and $SlotTime = 13$ μ. $SIFS$ is the acronym for *Short Interframe Space* and $AIFSN$ for *Arbitration Interframe Space Number*			

Table 3.15. *802.11p protocol medium access category parameters (IEEE 2012; ETSI 2020)*

3.6.3. *Modeling the propagation channel*

The propagation model utilized for all the simulations in Chapter 4 is the *Free Space* model proposed by default in *Riverbed Modeler*. However, OPNET models the transmission chain of wireless links as *pipelines* and gives the possibility to modify the default received power model of a WLAN network (Figure 3.27), as explained in Benferhat (2009).

Attributes	Value
Minimal frequency	5.855 GHz
Bandwidth	10 Mhz
Transmission power	0.005 W (i.e. 7 dBm)
Reception threshold	95 dBm
Transmission rate	12 Mbits/s
Maximal range	500 m

Table 3.16. *802.11p nodes configuration in OPNET*

```
/* Calculate wavelength (in meters).                          */
lambda = C / tx_center_freq;

/* Get distance between transmitter and receiver (in          */
/* meters).                                                   */
prop_distance = op_td_get_dbl (pkptr, OPC_TDA_RA_START_DIST);

/* Compute the path loss for this distance and                */
/* wavelength.                                                */
if (prop_distance > 0.0)
        {
        path_loss = (lambda * lambda) /
                (SIXTEEN_PI_SQ * prop_distance * prop_distance);
        } /* Free Space Model - SIXTEEN_PI_SQ is 16 times pi-squared */
else
        path_loss = 1.0;
}
```

Figure 3.27. *Free-space attenuation modeling – taken from the wlan_power.ps.c file of OPNET*

When the development of a protocol impacts the operation of radio technology, it is important to study the influence of the propagation channel on its performance, with in the case of VANETs attenuation models in urban or road environments. Examples of models include the LOS/OLOS models proposed by Abbas et al. (2015) or still the Nakagami models proposed in ETSI (2012).

The implementation of the CBL scheme only impacts the network layer and therefore cannot be optimized according to changes in the propagation channel. However, during the structural analysis of CBL, we wanted to evaluate its sensitivity (section 4.3.3) when direct neighboring vehicles are considered as obstacles to signal propagation to more distant vehicles (OLOS scenario). For this purpose, we used as a guideline the results of the studies by Abbas et al. (2015) on modeling LOS/OLOS in the highway context and implemented a very free and simplified implementation of the OLOS phenomenon. Figure 3.28 of Abbas et al. (2015) is the result of modeling measurements carried out at the 5.6 Ghz carrier frequency. This figure shows that for a maximal communication range of 500 m with LOS (Lund-LOS curve), the received power is 85 dBm. For the same received power, the Lund-OLOS curve shows a maximal communication range of 210 m. In order not to modify the OPNET parameterization of Table 3.16 (transmission power and reception threshold set to 0.005 W and 95 dBm, respectively), the problem of modeling LOS/OLOS was interpreted as follows. Below 210 m, propagation is done in free space according to the modeling proposed by default in OPNET, whatever the propagation configuration (LOS or OLOS). Above 210 m, the propagation configuration is verified by adopting a solution inspired by the work of Akhtar et al. (2015), which considers that "any vehicle that obstructs the [first] Fresnel ellipsoid might affect the transmitted signal". The Fresnel ellipsoid is a volume in space defined between the antenna of a transmitter and that of a receiver used to evaluate the electromagnetic wave propagation attenuation caused by an obstacle. We interpreted it in a simplified way as a 4-m wide rectangle between the source and receiving nodes (altitude has already been neglected in node trajectories, section 3.5.1). If a neighboring node lies within the rectangular area, a packet received by the node more than 210 m from the source node is considered unreadable: the receiving node rejects it at the CBL implementation level (Rivoirard 2018).

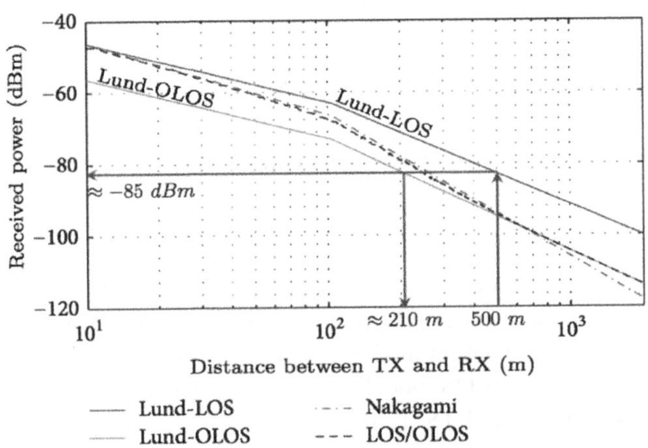

Figure 3.28. *Average power received for the LOS/OLOS model and the Nakagami model; figure by Abbas et al. (2015) and annotated by Rivoirard (2018). For a color version of this figure, see www.iste.co.uk/wahl/design.zip*

In summary, all the models in Chapter 4 except the study in section 4.3.3 consider a free-space propagation model. For the analysis of the structural performance of CBL-OLSR in LOS/OLOS compared to a free-space model, this simplified modeling of a LOS/OLOS model was used in section 4.3.3.

3.6.4. *Modeling the CBL-OLSR protocol in 802.11p nodes*

Figure 3.29 shows the process model of a WLAN mobile workstation-type node and the invocation location of OLSR and AODV child processes. The OLSR proactive routing protocol is invoked by the root process "manet_rte_mgr.pr.m" at the MANET route manager "manet_rte_ mgr" of *Riverbed Modeler*. For its part, the AODV reactive routing protocol is invoked at the "ip" level by the "manet_mgr" process, which in turn is invoked by the "ip_dispatch" process.

The CBL pro-active scheme was implemented in a copy of the "olsr_rte.pr.m" *Process model*, named "cbl-olsr_rte.pr.m". Consequently, an invocation of the CBL-OLSR protocol was added to the code of the "manet_rte_mgr.pr.m" root process at the MANET route manager level. By maintaining the same implementation structure as that of the OLSR process, the compatibility of the CBL-OLSR node with the native components was preserved. In particular, no *cross-layer* mechanism was implemented with the other layers. Therefore, the routing statistics obtained during the simulations presented in Chapter 4 are attributable only to the mechanisms implemented for routing.

The native OLSR protocol data structures that represent the neighbor tables were modified for the implementation of CBL-OLSR to incorporate the backup of additional information necessary for the operation of the CBL scheme. Consequently, a CBL-OLSR version of the "olsr_pkt_support.h" file was created that modifies and completes the set of data structures of HELLO message headers and packets with the information necessary to the CBL scheme (Rivoirard et al. 2020). These include those added to the "Message Type" mentioned in section 2.5.3.2 (Table 2.2) or still those related to speed, steering angle and vehicle position described in Rivoirard et al. (2020).

The changes made to the native OLSR process for implementing the CBL routing scheme are as follows. The function executed each time HELLO messages are received (Figure 2.26) and was modified to execute the CBL algorithm. The MPR selection heuristic was replaced by the branch node election procedure. New statistics and their associated computations were added. CBL own global variables, including time thresholds, were also introduced (section 2.5.2.5).

A version of the CBL-OLSR *Process model* that takes into account the contributions of the modified OLSR version presented in Figure 3.22b was also carried out in order to take into account the multi-trace trajectories (section 3.5.1.2.2) necessary for modeling the scenarios of networks R1 to R3 simulated over a time period of 500 s.

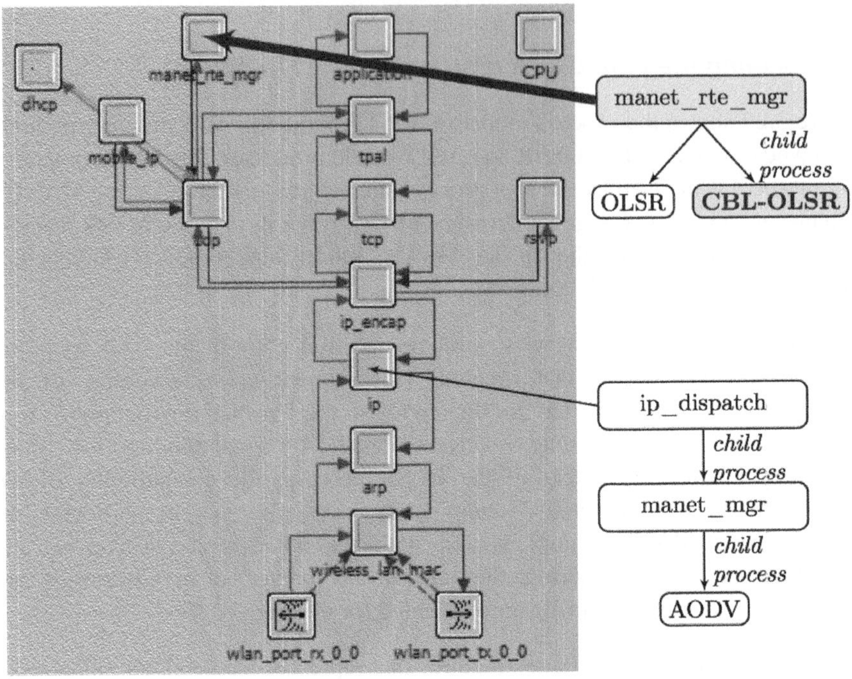

Figure 3.29. *802.11p mobile workstation-type node process model and invocation location of the CBL-OLSR child process . For a color version of this figure, see www.iste.co.uk/wahl/design.zip*

3.7. Conclusion

Important developments in terms of tools and measurements have been achieved with the advent of studies on MANETs, and more particularly since the 2000s with the momentum given to VANETs as a complement to the communication infrastructure in ITSs. The purpose of this chapter was to highlight the challenges created by the modeling and simulation of a routing protocol for VANETs and to provide some answers, given the knowledge and technologies of the moment. In particular, the specific issue of modeling the mobility of a large number of vehicles through mobile communication nodes was addressed and development devices were proposed. Another objective of this chapter was the presentation of the mobility scenarios, the implementation of the CBL-OLSR case study protocol and the parameterization of the IEEE 802.11p technology of mobile nodes that are used in Chapter 4, and are dedicated to evaluations.

A general observation that can be retrieved from this presentation is that the tools for assessing communication networks for vehicles and land transport in general are constantly evolving. In fact, it appears that this evolution has increasingly accelerated in recent years with the success of simulation and virtual laboratories as a means of accelerating evaluations at a lower cost; this is particularly the case when considering the plethora of propositions for routing protocols, technologies, scenarios and applications to be evaluated to demonstrate the relevance of a set of choices in the context of a particular deployment.

The availability and sharing of a large amount of data on mobility and the use of communication networks in many parts of the world, down to a very detailed level for certain cities, has greatly contributed to improve the means of evaluation, in particular, through micro–meso–macro-mobility simulators that complement communication network simulators, thus allowing evaluations on increasingly realistic and therefore more relevant scenarios.

3.8. References

3GPP (2023). The 5G standard. 3GPP specifications & technologies releases [Online]. Available at: https://www.3gpp.org/specifications-technologies/releases [Accessed 17 April 2023].

Abbas, T., Sjöberg, K., Karedal, J., Tufvesson, F. (2015). A measurement based shadow fading model for vehicle-to-vehicle network simulations. *International Journal of Antennas and Propagation*, 190607, 12. doi: 10.1155/2015/190607.

Abdelgader, A. and Lenan, W. (2014). The physical layer of the IEEE 802.11 p WAVE communication standard: The specifications and challenges. In *Proceedings of the World Congress on Engineering and Computer Science, Volume 2*, 22–24 October, San Francisco.

Ahmed, B., Malik, A.W., Hafeez, T., Ahmed, N. (2019). Services and simulation frameworks for vehicular cloud computing: A contemporary survey. *EURASIP Journal on Wireless Communications and Networking*, 4, 21, Busan, 133–138. doi: 10.1186/s13638-018-1315-y.

Akhtar, N. Ergen, S.C., Ozkasap, O. (2015). Vehicle mobility and communication channel models for realistic and efficient highway VANET simulation. *IEEE Transactions on Vehicular Technology*, 64(1), 248–262. doi: 10.1109/ TVT.2014.2319107.

Al-Sultan, S., Al-Doori, M.M., Al-Bayatti, A.H., Zedan, H. (2014). A comprehensive survey on vehicular ad hoc network. *Journal of Network and Computer Applications*, 37, 380–392. doi: 10.1016/j.jnca.2013.02.036.

Amjad, Z. and Song, W.-C. (2015). Road aware QoS routing in VANETs. In *2015 17th Asia-Pacific Network Operations and Management Symposium (APNOMS)*, Busan, 133–138. doi: 10.1109/APNOMS.2015.7275415.

Amoozadeh, M. (2017). VENTOS – VEhicular NeTwork Open Simulator. Rubinet Lab, ECE Department University of California, Davis [Online]. Available at: https://maniam.github.io/VENTOS/ [Accessed 25 June 2021].

Amoozadeh, M., Deng, H., Chuah, C.-N., Zhang, H.M., Ghosal, D. (2015). Platoon management with cooperative adaptive cruise control enabled by VANET. *Vehicular Communications*, 2(2), 110–123. doi: 10.1016/j.vehcom. 2015.03.004.

Amoozadeh, M., Ching, B., Chuah, C.-N., Ghosal, D., Zhang, H.M. (2019). Ventos: Vehicular network open simulator with hardware-in-the-loop support. *Procedia Computer Science*, 151, 61–68. doi: 10.1016/j.procs.2019.04.012.

Arcep (2019). Réseaux du futur – note n°2 les voitures connectées. Autorité de régulation des communications électroniques et des postes de la République Française [Online]. Available at: https://www.arcep.fr/ [Accessed 4 June 2021].

Arkian, H.R., Atani, R.E., Pourkhalili, A., Kamali, S. (2015). A stable clustering scheme based on adaptive multiple metric in vehicular ad-hoc networks. *Journal of Information Science & Engineering*, 31(2), 361–386.

Ashtaiwi, A., Altayesh, A., Belghet, K. (2015). IEEE 802.11 p performance evaluation at different driving environments. In *2015 World Symposium on Computer Networks and Information Security (WSCNIS)*, Hammamet, 1–8. doi: 10.1109/wscnis. 2015.7368286.

Azizian, M., Cherkaoui, S., Hafid, A.S. (2016). DCEV: A distributed cluster formation for VANET based on end-to-end realtive mobility. In *2016 International Wireless Communications and Mobile Computing Conference (IWCMC)*. IEEE, Paphos, 287–291. doi: 10.1109/iwcmc.2016.7577072.

Bai, F., Sadagopan, N., Helmy, A. (2003). IMPORTANT: A framework to systematically analyze the impact of mobility on performance of RouTing protocols for Adhoc NeTworks. In *IEEE INFOCOM 2003. Twenty-Second Annual Joint Conference of the IEEE Computer and Communications Societies (IEEE Cat. No.03CH37428), Volume 2*, San Francisco, 825–835. doi: 10.1109/infcom.2003.1208920.

Benferhat, S. (2009). Simulation des conditions de trafic intracellulaire d'un réseau sans fil en milieu industriel par un modèle de propagation composite. PhD Thesis, Université Blaise Pascal, Clermont-Ferrand [Online]. Available at: https://www.theses.fr/2009CLF21923.

Bilstrup, K., Uhlemann, E., Ström, E.G. (2008). Medium access control in vehicular networks based on the upcoming IEEE 802.11 p standard. In *15th World Congress on Intelligent Transport Systems (ITS)*, 15–20 November. Jacob K. Javits Convention Center, World Congress on ITS, New York.

Bitam, S. and Mellouk, A. (2011). QoS swarm bee routing protocol for vehicular ad hoc networks. In *2011 IEEE International Conference on Communications (ICC)*, Kyoto 1–5. doi: 10.1109/icc.2011.5963424.

Boban, M., Vinhoza, T.T., Ferreira, M., Barros, J., Tonguz, O.K. (2011). Impact of vehicles as obstacles in vehicular ad hoc networks. *IEEE Journal on Selected Areas in Communications*, 29(1), 15–28. doi: 10.1109/JSAC.2011.110103.

Buisson, C., Villegas, D., Rivoirard, L. (2016). Using polar coordinates to filter trajectories data without adding extra physical constraints. In *Transportation Research Board 95th Annual Meeting* [Online]. Available at: https://trid.trb.org/view/1393268.

C-Roads (2023). Projet CEF France InDiD – Infrastructure digitale de demain [Online]. Available at: https://www.c-roads.eu/pilots/core-members/france/Partner/project/show/indid.html.

Camp, T., Boleng, J., Davies, V. (2002). A survey of mobility models for ad hoc network research. *Wireless Communications and Mobile Computing*, 2(5), 483–502. doi: 10.1002/wcm.72.

Cavalcanti, E.R., de Souza, J.A.R., Spohn, M.A., de Morais Gomes, R.C., da Costa, A.F.B.F. (2018). VANETs' research over the past decade: Overview, credibility, and trends. *SIGCOMM Comput. Commun. Rev.*, 48(2), 31–39. doi: 10.1145/3213232.3213237.

Cha, S.-H., Lee, J.-E., Ryu, M. (2016). Directed broadcasting with mobility prediction for vehicular sensor networks. *International Journal of Distributed Sensor Networks*, 12(7). doi: 10.1177/1550147716657930.

Costeseque, G. (2011). Modélisation du trafic routier : passage du microscopique au macroscopique. Master's Thesis, Université de Lyon, Lyon [Online]. Available at: http://www-sop.inria.fr/members/Guillaume.Costeseque/premi_ abertis_costeseque.pdf [Accessed 17 February 2021].

Crawdad (2020a). CRAWDAD, A Community Resource for Archiving Wireless Data At Dartmouth, FAQ [Online]. Available at: https://ieee-dataport.org/help/crawdad-faq [Accessed 1 December 2023].

Crawdad (2020b). CRAWDAD library – a public group library of papers about CRAWDAD datasets or about the use of CRAWDAD datasets [Online]. Available at: https://www.zotero.org/groups/2614522/crawdad/ [Accessed 2 February 2021].

Daeinabi, A., Rahbar, A.G.P., Khademzadeh, A. (2011). VWCA: An efficient clustering algorithm in vehicular ad hoc networks. *Journal of Network and Computer Applications*, 34(1), 207–222. doi: 10.1016/j.jnca.2010.07.016.

Daknou, E., Thaalbi, M., Tabbane, N. (2015). A fast clustering algorithm for VANETs. In *Proceedings of the 13th International Conference on Advances in Mobile Computing and Multimedia*. ACM, 195–202. doi: 10.1145/2837126.2837147.

Dar, K., Bakhouya, M., Gaber, J., Wack, M., Lorenz, P. (2010). Wireless communication technologies for ITS applications [topics in automotive networking]. *IEEE Communications Magazine*, 48(5), 156–162. doi: 10.1109/mcom.2010.5458377.

DCAITI (2021). Welcome to eclipse MOSAIC download area – Eclipse MOSAIC: VSimRTI, but open-source [Online]. Available at: https://www.dcaiti.tu-berlin.de/research/simulation/download/ [Accessed 19 January 2021].

Dhoutaut, D., Régis, A., Spies, F. (2006). Impact of radio propagation models in vehicular ad hoc networks simulations. In *VANET'06: Proceedings of the 3rd International Workshop on Vehicular Ad Hoc Networks*. ACM. doi: 10.1145/1161064.1161072.

DLR (2021a). Sumo documentation on the netconvert road network import and conversion tool [Online]. Available at: https://sumo.dlr.de/docs/netconvert.html.

DLR (2021b). Documentation on the netedit road network graphics editor included in SUMO [Online]. Available at: https://sumo.dlr.de/docs/Netedit/.

Eclipse (2021). Simple network simulator (SNS): Eclipse MOSAIC – A multi-domain and multi-scale simulation framework for connected and automated mobility [Online]. Available at: https://www.eclipse.org/mosaic/docs/simulators/network_simulator_sns/ [Accessed 1 February 2021].

EstiNet (2020). Estinet network simulator and emulator. EstiNetTechnologies [Online]. Available at: http://www.estinet.com/ns/ [Accessed 18 January 2021].

ETSI (2012). Intelligent transport systems (ITS); STDMA recommended parameters and settings for cooperative ITS: Access layer part. Technical Report ETSI TR 102 861 V1.1.1 (2012-01), ETSI Technical Committee Intelligent Transport System, Sophia Antipolis [Online]. Available at: http://www. etsi.org/deliver/etsi_tr/102800_102899/102861/01.01.01_60/tr_102861v010101p.pdf.

ETSI (2020). Intelligent Transport Systems (ITS): Access layer specification for Intelligent Transport Systems operating in the 5 GHz frequency band. European Standard ETSI EN 302 663 V1.3.1 (2020-01), ETSI Technical Committee Intelligent Transport System, Sophia Antipolis [Online]. Available at: https://www.etsi.org/deliver/etsi_en/302600_302699/302663/01.03.01_60/en_302663v010301p.pdf.

Fallgren, M., Dillinger, M., Alonso-Zarate, J., Boban, M., Abbas, T., Manolakis, K., Mahmoodi, T., Svensson, T., Laya, A., Vilalta, R. (2018). Fifth-generation technologies for the connected car: Capable systems for vehicle-to-anything communications. *IEEE Vehicular Technology Magazine*, 13(3), 28–38. doi: 10.1109/ MVT.2018.2848400.

Ferreira, M., Conceiçao, H., Fernandes, R., Tonguz, O.K. (2009). Urban connectivity analysis of VANETs through stereoscopic aerial photography. In *2009 IEEE 70th Vehicular Technology Conference Fall*, Anchorage, 1–3. doi: 10.1109/vetecf.2009.5378962.

FHWA (2020). Traffic analysis tools: Next generation simulation (NGSIM) - FHWA operations [Online]. Available at: https://ops.fhwa.dot.gov/trafficanalysistools/ngsim.htm [Accessed 2 February 2021].

Fiore, M. (2006). Mobility models in inter-vehicle communications literature. Report, Politecnico di Torino, Turin, 147.

Fiore, M., Harri, J., Filali, F., Bonnet, C. (2007). Vehicular mobility simulation for VANETs. *40th Annual Simulation Symposium (ANSS'07)*, Norfolk, VA. doi: 10.1109/ANSS.2007.44.

Fishman, G.S. (2001). *Discrete-Event Simulation: Modeling, Programming, and Analysis*. Springer, Berlin.

Fraunhofer FOKUS (2020). Fraunhofer FOKUS, ASCT (Automotive Services and Communication Technologies) simulation & testing – Smart mobility simulation [Online]. Available at: https://www.fokus.fraunhofer.de/en/asct/simulation_testing [Accessed 19 January 2021].

Garcia, M.H.C., Molina-Galan, A., Boban, M., Gozalvez, J., Coll-Perales, B., Şahin, T., Kousaridas, A. (2021). A tutorial on 5G NR V2X communications. *IEEE Communications Surveys Tutorials*, 23(3), 1972–2026. doi: 10.1109/COMST.2021.3057017.

GRC (2009). CityMob: Generador de patrones de movimiento para entornos VANET [Online]. Available at: http://www.grc.upv.es/Software/oldsw/city-mob.html [Accessed 26 February 2021].

Gruet, C. (2018). 5G état de l'art. In *Support de cours de la Formation Telecom Evolution*. Institut Mines Telecom, Paris.

Hafeez, K.A., Zhao, L., Liao, Z., Ma, B.N.-W. (2009). The optimal radio propagation model in VANET. In *2009 Fourth International Conference on Systems and Networks Communications*. IEEE. doi: 10.1109/icsnc.2009.101.

Halati, A., Lieu, H., Walker, S. (1997). CORSIM – corridor traffic simulation model. In *Traffic Congestion and Traffic Safety in the 21st Century: Challenges, Innovations, and Opportunities*, Benekohal, R.F. et al. (eds). American Society of Civil Engineers, New York [Online]. Available at: http://cedb.asce.org/CEDBsearch/record.jsp?dockey=0105676.

Harrabi, S., Jaffar, I.B., Ghedira, K. (2016). Novel optimized routing scheme for VANETs. *Procedia Computer Science*, 98, 32–39. doi: 10.1016/ j.procs.2016.09.008.

Härri, J., Filali, F., Bonnet, C. (2006a). Mobility models for vehicular ad hoc networks: A survey and taxonomy. Research report RR-06-168, EURECOM, 33 [Online]. Available at: https://www.eurecom.fr/publication/1951 [Accessed 4 February 2021].

Härri, J., Filali, F., Bonnet, C., Fiore, M. (2006b). VanetMobiSim: Generating realistic mobility patterns for VANETs. *Proceedings of the 3rd International Workshop on Vehicular Ad Hoc Networks*. doi: 10.1145/1161064.1161084.

Härri, J., Filali, F., Bonnet, C. (2009). Mobility models for vehicular ad hoc networks: A survey and taxonomy. *IEEE Communications Surveys Tutorials*, 11(4), 19–41. doi: 10.1109/ SURV.2009.090403.

He, G. (2002). Destination-sequenced distance vector (DSDV) protocol. Networking Laboratory, Helsinki University of Technology. doi: 10.1109/icfn.2009.51.

Hnatyshin, V. and Asenov, H. (2010). Design and implementation of an OPNET model for simulating GeoAODV MANET routing protocol. In *Proc. of the OPNETWORK 2010 International Conference, Session: Wireless Ad Hoc and Wireless Personal Area Networks*. OPNET Technologies, Inc., Washington [Online]. Available at: https://api. semanticscholar.org/CorpusID:18625228.

Hoogendoorn, S., Van Zuylen, H., Schreuder, M., Gorte, B., Vosselman, G. (2003). Microscopic traffic data collection by remote sensing. *Transportation Research Record: Journal of the Transportation Research Board*, 1855, 121–128. doi: 10.3141/1855-15.

Hussey, R., Huff, E., Shinwari, Z., Hnatyshin, V. (2013). A comparative study of proactive and reactive geographical routing protocols for MANET. In *Proc. of the International Conference on Wireless Networks (ICWN)*, Csrea, London.

IEEE (2010). IEEE 802.11p-2010 – IEEE standard for information technology – Local and metropolitan area networks – Specific requirements – Part 11: Wireless LAN medium access control (MAC) and physical layer (PHY) specifications amendment 6: Wireless access in vehicular environments [Online]. Available at: https://standards.ieee.org/ standard/802_11p-2010.html.

IEEE (2012). IEEE standard for information technology – Telecommunications and information exchange between systems local and metropolitan area networks – Specific requirements – Part 11: Wireless LAN medium access control (MAC) and physical layer (PHY) specifications. IEEE Std 802.11-2012 (Revision of IEEE Std 802.11-2007).

InterCor (2019). European ERTICO InterCor (Interoperable Corridors) Project 2017–February 2020 [Online]. Available at: http://intercor-project.eu/ [Accessed 18 January 2021].

Johnson, D. and Maltz, D. (1996). Dynamic source routing in ad hoc wireless networks. In *Mobile Computing*, Imielinski, T. and Korth, H. (eds). Springer, Boston, 153–181. doi: 10.1007/978-0-585-29603-6l_5.

Karedal, J., Czink, N., Paier, A., Tufvesson, F., Molisch, A.F. (2011). Path loss modeling for vehicle-to-vehicle communications. *IEEE Transactions on Vehicular Technology*, 60(1), 323–328. doi: 10.1109/tvt.2010.2094632.

Krajzewicz, D., Hertkorn, G., Rössel, C., Wagner, P. (2010). SUMO (Simulation of Urban MObility) – An open-source traffic simulation. In *Proceedings of the 4th Middle East Symposium on Simulation and Modelling (MESM20002)*, 183–187 [Online]. Available at: https://elib.dlr.de/6661/2/dkrajzew_MESM2002.pdf [Accessed 20 March 2024].

Krauß, S. (1998). Microscopic modeling of traffic flow: Investigation of collision free vehicle dynamics. PhD Thesis, Universität zu Köln, Cologne [Online]. Available at: https://sumo.dlr.de/pdf/KraussDiss.pdf.

Kumar, J., Singh, A., Panda, M., Bhadauria, H. (2016). Study and performance analysis of routing protocol based on CBR. *Procedia Computer Science*, 85, 23–30. doi: 10.1016/j.procs.2016.05.172.

Lan, K. (2009). MOVE: A practical simulator for mobility model in VANET. In *Telematics Communication Technologies and Vehicular Networks*. IGI Global. doi: 10.4018/978-1-60566-840-6.ch021.

Li, Y., Jin, D., Wang, Z., Zeng, L., Chen, S. (2013). Exponential and power law distribution of contact duration in urban vehicular ad hoc networks. *IEEE Signal Processing Letters*, 20(1), 110–113. doi: 10.1109/lsp.2012.2231412.

Li, G., Boukhatem, L., Martin, S. (2015). An intersection-based QoS routing in vehicular ad hoc networks. *Mobile Networks and Applications*, 20(2), 268–284. doi: 10.1007/s11036-015-0577-4.

Lin, Y.-W. and Huang, G.-T. (2012). Optimal next hop selection for VANET routing. In *7th International Conference on Communications and Networking in China*. IEEE, Kunming, 611–615. doi: 10.1109/chinacom.2012.6417556.

Lopez, P.A., Behrisch, M., Bieker-Walz, L., Erdmann, J., Flötteröd, Y.-P., Hilbrich, R., Lücken, L., Rummel, J., Wagner, P., Wießner, E. (2018). Microscopic traffic simulation using SUMO. In *The 21st IEEE International Conference on Intelligent Transportation Systems*, Maui, 2575–2582. doi: 10.1109/ITSC.2018.8569938.

Mangel, T., Michl, M., Klemp, O., Hartenstein, H. (eds) (2011). Real-world measurements of non-line-of-sight reception quality for 5.9 GHz IEEE 802.11 p at intersections. In *International Workshop on Communication Technologies for Vehicles*. Springer, Berlin. doi: 10.1007/978-3-642-19786-4_17.

Mantoro, T. and Reza, M. (2016). Performance analysis of AODV and DSDV using SUMO, MOVE and NS2. In *2016 International Conference on Informatics and Computing (ICIC)*, Mataram, 372–376. doi: 10.1109/IAC.2016.7905747.

Martinez, F.J., Cano, J.-C., Calafate, C.T., Manzoni, P. (2008). CityMob: A mobility model pattern generator for VANETs. In *ICC Workshops – 2008 IEEE International Conference on Communications Workshops*, Beijing, 370–274. doi: 10.1109/ICCW.2008.76.

Martinez, F.J., Fogue, M., Coll, M., Cano, J.-C., Calafate, C.T., Manzoni, P. (2010). Assessing the impact of a realistic radio propagation model on VANET scenarios using real maps. In *2010 Ninth IEEE International Symposium on Network Computing and Applications*, Cambridge, MA, 132–139. doi: 10.1109/nca.2010.24.

Martinez, F.J., Toh, C.K., Cano, J.-C., Calafate, C.T., Manzoni, P. (2011). A survey and comparative study of simulators for vehicular ad hoc networks (VANETs). *Wireless Communications and Mobile Computing*, 11(7), 813–828. doi: 10.1002/wcm.859.

Maslekar, N., Boussedjra, M., Mouzna, J., Labiod, H. (2011). C-DRIVE: Clustering based on direction in vehicular environment. In *2011 4th IFIP International Conference on New Technologies, Mobility and Security*. IEEE, Baden-Baden, 845–850. doi: 10.1109/NTMS.2011.5720653.

MathWorks (2021). Matlab documentation [Online]. Available at: https://www.mathworks.com/help/matlab/. [Accessed 18 January 2021].

Meireles, R., Boban, M., Steenkiste, P., Tonguz, O., Barros, J. (2010). Experimental study on the impact of vehicular obstructions in VANETs. In *2010 IEEE Vehicular Networking Conference*. Jersey City, 338–345. doi: 10.1109/vnc.2010.5698233.

Mohapatra, S. and Kanungo, P. (2012). Performance analysis of AODV, DSR, OLSR and DSDV routing protocols using NS2 simulator. *Procedia Engineering (International Conference on Communication Technology and System Design 2011)*, 30, 69–76. doi: 10.1016/j.proeng.2012.01.835.

Molina-Masegosa, R., Gozalvez, J., Sepulcre, M. (2020). Comparison of IEEE 802.11p and LTE-V2X: An evaluation with periodic and aperiodic messages of constant and variable size. *IEEE Access*, 8, 121526–121548. doi: 10.1109/ACCESS.2020.3007115.

Molisch, A.F., Tufvesson, F., Karedal, J., Mecklenbrauker, C.F. (2009). A survey on vehicle-to-vehicle propagation channels. *IEEE Wireless Communications*, 16(6), 12–22. doi: 10.1109/mwc.2009. 5361174.

Mukunthan, A. (2015). Routing and propagation in urban vehicular ad hoc networks. PhD Thesis, University of Wollongong, Wollongong [Online]. Available at: http://ro.uow.edu.au/theses/4558/.

Naik, G., Choudhury, B., Park, J.-M. (2019). IEEE 802.11bd & 5G NR V2X: Evolution of radio access technologies for V2X communications. *IEEE Access*, 7, 70169–70184. doi: 10.1109/ACCESS.2019.2919489.

nsnam (2014). ns-2 Network Simulator [Online]. Available at: http://nsnam.sourceforge.net/wiki/index.php/Main_Page year=2014 [Accessed 18 January 2021].

nsnam (2021). ns-3 | a discrete-event network simulator for internet systems [Online]. Available at: https://www.nsnam.org/wiki/Main_Page [Accessed 18 January 2021].

Ollinger, E. (2020). Projet SCOOP : véhicules et routes connectés [Online]. Available at: http://www.scoop.developpement-durable.gouv.fr/ [Accessed 18 January 2021].

Omnetpp (2010). VSimRTI – Smart mobility simulation [Online]. Available at: https://omnetpp.org/download-items/VSimRTI.html [Accessed 19 January 2021].

OpenSimLtd (2019a). Developer's guide – INET 4.3.0 [Online]. Available at: https://inet.omnetpp.org/docs/developers-guide/index.html [Accessed 3 June 2021].

OpenSimLtd (2019b). OMNeT++ simulation environment [Online]. Available at: https://omnetpp.org/intro/ [Accessed 18 January 2021].

OpenSimLtd (2019c). User's guide – INET 4.3.0 [Online]. Available at: https://inet.omnetpp.org/docs/users-guide/index.html [Accessed 3 June 2021].

OpenStreetMap (2021). OpenStreetMap France – La carte coopérative libre [Online]. Available at: https://www.openstreetmap.fr/.

Patel, M.K. (2012). Comparative study of vehicular ad-hoc network mobility models and simulators. *International Journal of Computer Applications*, 47(6), 38–43. doi: 10.5120/7196-9967.

Protzmann, R., Schünemann, B., Radusch, I. (eds) (2017). Simulation of convergent networks for intelligent transport systems with VSimRTI. In *Networking Simulation for Intelligent Transportation Systems*, Hilt, B., Berbineau, M., Vinel, A., Pirovano, A. (eds). ISTE Ltd, London, and John Wiley & Sons, New York. doi: 10.1002/ 9781119407447.ch1.

Rawashdeh, Z.Y. and Mahmud, S.M. (2012). A novel algorithm to form stable clusters in vehicular ad hoc networks on highways. *EURASIP Journal on Wireless Communications and Networking*, 1, 15. doi: 10.1186/1687-1499-2012-15.

Ren, M., Khoukhi, L., Labiod, H., Zhang, J., Vèque, V. (2017). A mobility-based scheme for dynamic clustering in vehicular ad-hoc networks (VANETs). *Vehicular Communications*, 9, 233–241. doi: 10.1016/j.vehcom.2016.12.003.

Rhattoy, A. and Zatni, A. (2012). The impact of propagation environment and traffic load on the performance of routing protocols in ad hoc networks. *International Journal of Distributed and Parallel Systems (IJDPS)*, 3(1), 75–87. doi: 10.5121/ijdps.2012.3106.

Riverbed (2021). Riverbed modeler (formerly referred to as OPNET modeler suite) [Online]. Available at: https://support.riverbed.com/content/support/software/opnet-model/modelerwireless-suite.html [Accessed 18 January 2021].

Rivoirard, L. (2018). Modèle d'auto-organisation pour les protocoles de routage dans les réseaux ad hoc de véhicules: application à la perception élargie et à la localisation coopératives. PhD Thesis, Université de Lille, Lille [Online]. Available at: http://www.theses.fr/2018LIL1I041.

Rivoirard, L., Wahl, M., Sondi, P., Berbineau, M., Gruyer, D. (2016). Using real-world car traffic dataset in vehicular ad hoc network performance evaluation. *International Journal of Advanced Computer Science and Applications (IJACSA)*, 7(12), 390–398. doi: 10.14569/IJACSA.2016.071251.

Rivoirard, L., Wahl, M., Sondi, P., Berbineau, M., Gruyer, D. (2017). Performance evaluation of AODV, DSR, GRP and OLSR for VANET with real-world trajectories. In *15th International Conference on ITS Telecommunications (ITST)*. IEEE, Warsaw, 1–7. doi: 10.1109/itst.2017.7972224.

Rivoirard, L., Wahl, M., Sondi, P., Berbineau, M., Gruyer, D. (2018). Chain-Branch-Leaf: A clustering scheme for vehicular networks using only V2V communications. *Ad Hoc Networks*, 68, 70–84. doi: 10.1016/j.adhoc.2017.10.007.

Rivoirard, L., Wahl, M., Sondi, P. (2020). Multipoint relaying versus chain-branch-leaf clustering performance in optimized link state routing-based vehicular ad hoc networks. *IEEE Transactions on Intelligent Transportation Systems*, 21(3), 1034–1043. doi: 10.1109/TITS.2019.2900767.

Roy, A., Paul, B., Paul, S.K. (2015). VANET topology based routing protocols & performance of AODV, DSR routing protocols in random waypoint scenarios. In *2015 International Conference on Computer and Information Engineering (ICCIE)*. IEEE, Rajshahi, 50–53. doi: 10.1109/ccie.2015.7399315.

Sarkar, T.K., Zhong, J., Kyungjung, K., Medouri, A., Salazar-Palma, M. (2003). A survey of various propagation models for mobile communication. *IEEE Antennas and Propagation Magazine*, 45(3), 51–82. doi: 10.1109/MAP.2003.1232163.

Scalable (2021). QualNet – Network simulation software | SCALABLE Networks [Online]. Available at: https://www.scalable-networks.com/products/qualnet-network-simulation software-tool/ [Accessed 26 January 2021].

Sehla, K., Nguyen, T.M.T., Pujolle, G., Velloso, P.B. (2022). Resource allocation modes in C-V2X: From LTE-V2X to 5G-V2X. *IEEE Internet of Things Journal*, 9(11), 8291–8314. doi: 10.1109/JIOT.2022.3159591.

Sethi, V. and Chand, N. (2017). A destination based routing protocol for context based clusters in VANET. *Communications and Network*, 9(3), 179. doi: 10.4236/cn.2017.93013.

Singh, P.K. and Lego, K. (2011). Comparative study of radio propagation and mobility models in vehicular ad hoc network. *International Journal of Computer Applications*, 16(8), 37–42. doi: 10.5120/2031-2600.

Singh, P.K., Nandi, S.K., Nandi, S. (2019). A tutorial survey on vehicular communication state of the art, and future research directions. *Vehicular Communications*, 18, 39. doi: 10.1016/j.veh-com.2019.100164.

Sommer, C. (2020). Veins – vehicles in network simulation. The open source vehicular network simulation framework [Online]. Available at: https://veins.car2x.org/ [Accessed 19 January 2021].

Sommer, C. and Dressler, F. (2008). Progressing toward realistic mobility models in VANET simulations. *IEEE Communications Magazine*, 46(11), 132–137. doi: 10.1109/MCOM.2008.4689256.

Sommer, C., German, R., Dressler, F. (2011). Bidirectionally coupled network and road traffic simulation for improved IVC analysis. *IEEE Transactions on Mobile Computing*, 10(1), 3–15. doi: 10.1109/TMC.2010.133.

Sommer, C., Eckhoff, D., Brummer, A., Buse, D.S., Hagenauer, F., Joerer, S., Segata, M. (2019). Veins – The open source vehicular network simulation framework. In *Recent Advances in Network Simulation*, Virdis, A. and Kirsche, M. (eds). Springer, Berlin.

Sondi, P. (2010). Le routage à qualité de service dans les réseaux mobiles ad hoc (Routing with quality of service support in mobile ad hoc networks). PhD Thesis, Université Polytechnique Hauts-de-France, Valenciennes [Online]. Available at: https://tel.archives-ouvertes.fr/tel-00551243.

Sondi, P., Gantsou, D., Lecomte, S. (2010). Performance evaluation of multimedia applications over an OLSR-based mobile ad hoc network using OPNET. In *12th International Conference on Computer Modelling and Simulation (UKSim)*. IEEE, Cambridge, 567–572. doi: 10.1109/uksim.2010.109.

Sondi, P., Gantsou, D., Lecomte, S. (2013). Design guidelines for quality of service support in optimized link state routing-based mobile ad hoc networks. *Ad Hoc Networks*, 11(1), 298–323. doi: 10.1016/j.adhoc.2012.06.001.

Sondi, P., Wahl, M., Rivoirard, L., Cohin, O. (2016). Performance evaluation of 802.11 p-based ad hoc vehicle-to-vehicle communications for usual applications under realistic urban mobility. *International Journal of Advanced Computer Science and Applications*, 7(5), 221–230. doi: 10.14569/ijacsa.2016.070532.

Sun, Y., Luo, S., Dai, Q., Ji, Y. (2015). An adaptive routing protocol based on QoS and vehicular density in urban VANETs. *International Journal of Distributed Sensor Networks*, 11(6), 631092. doi: 10.1155/2015/631092.

Teramoto, E., Baba, M., Mori, H., Asano, Y., Morita, H. (1997). NETSTREAM: Traffic simulator for evaluating traffic information systems. In *Proceedings of Conference on Intelligent Transportation Systems*. IEEE, Boston, MA, 484–489. doi: 10.1109/itsc.1997.660522.

Tian, D., Wang, Y., Lu, G., Yu, G. (2010). A VANETs routing algorithm based on Euclidean distance clustering. In *2010 2nd International Conference on Future Computer and Communication*, Wuhan, V1-183–V1-187. doi: 10.1109/ICFCC.2010.5497808.

Tripp-Barba, C., Urquiza-Aguiar, L., Zaldívar-Colado, A., Estrada-Jiménez, J., Aguilar-Calderón, J.A., Igartua, M.A. (2018). Comparison of propagation and packet error models in vehicular networks performance. *Vehicular Communications*, 12, 1–13.

Viriyasitavat, W., Boban, M., Tsai, H.-M., Vasilakos, A. (2015). Vehicular communications: Survey and challenges of channel and propagation models. *IEEE Vehicular Technology Magazine*, 10(2), 55–66. doi: 10.1109/mvt.2015.2410341.

Wahab, O.A., Otrok, H., Mourad, A. (2013). VANET QoS-OLSR: QoS-based clustering protocol for vehicular ad hoc networks. *Computer Communications*, 36(13), 1422–1435. doi: 10.1016/j.comcom.2013.07.003.

Wahl, M., Sondi, P., Rivoirard, L. (2021). Enhanced CBL clustering performance versus GRP, OLSR and AODV in vehicular ad hoc networks. *Telecommunication Systems*, 76, 525–540. doi: 10.1007/s11235-020-00734-1.

Wang, S.-Y. (2013). NCTU network simulator and emulator [Online]. Available at: http://nsl.cs.nctu.edu.tw/NSL/nctuns.html [Accessed 18 January 2021].

Wang, L. (2020). VANET Toolbox: A vehicular network simulator based on DES. MATLAB central file exchange [Online]. Available at: https://www.mathworks.com/matlabcentral/fileexchange/68437-vanet-toolboxa-vehicular-network-simulator-based-on-des [Accessed 3 June 2021].

Wang, S.-Y. and Lin, C.-C. (2008). NCTUns 5.0: A network simulator for IEEE 802.11(p) and 1609 wireless vehicular network researches. In *2008 IEEE 68th Vehicular Technology Conference*, Calgary, 1–2. doi: 10.1109/vetecf.2008.464.

Wang, S.-Y., Chiu, H.-W., Zheng, J., Chen, X. (2013). A cooperative approach to fully utilizing the aggregate bandwidth of all service channels in IEEE 802.11p/1609 networks. In *2013 IEEE Symposium on Computers and Communications (ISCC)*, Split, 000354–000359. doi: 10.1109/ISCC.2013.6754972.

Wang, L., Iida, R., Wyglinski, A.M. (2019). Vehicular network simulation environment via discrete event system modeling. *IEEE Access*, 7, 87246–87264. doi: 10.1109/ACCESS.2019.2922766.

Weber, J.S., Neves, M., Ferreto, T. (2021). VANET simulators: An updated review. *Journal of the Brazilian Computer Society*, 27(8), 1–12. doi: 10.1186/s13173-021-00113-x.

Wu, C., Ohzahata, S., Kato, T. (2013a). Can we generate efficient routes by using only beacons? Backbone routing in VANETs. In *2013 IEEE 24th Annual International Symposium on Personal, Indoor, and Mobile Radio Communications (PIMRC)*, London, 2929–2934. doi: 10.1109/pimrc.2013.6666648.

Wu, C., Ohzahata, S., Kato, T. (2013b). Data dissemination with dynamic backbone selection in vehicular ad hoc networks. In *2013 IEEE 78th Vehicular Technology Conference (VTC Fall)*, Las Vegas, 1–6. doi: 10.1109/vtcfall.2013.6692152.

Wu, C., Ohzahata, S., Ji, Y., Kato, T. (2016). How to utilize interflow network coding in VANETs: A backbone-based approach. *IEEE Transactions on Intelligent Transportation Systems*, 17(8), 2223–2237. doi: 10.1109/tits.2016.2516027.

Yacheur, B., Ahmed, T., Mosbah, M. (2020). Analysis and comparison of IEEE 802.11p and IEEE 802.11bd. *Communication Technologies for Vehicles: 15th International Workshop, Nets4Cars/Nets4Trains/Nets4Aircraft 2020*, 12574, 55–65. doi: 10.1007/978-3-030-66030-7_5.

Zeng, X., Bagrodia, R., Gerla, M. (1998). GloMoSim: A library for parallel simulation of large-scale wireless networks. In *Proceedings. Twelfth Workshop on Parallel and Distributed Simulation PADS '98 (Cat. No.98TB100233)*. IEEE, Banff, Alberta. doi: 10.1109/PADS.1998.685281.

Zheng, J., Wu, Y., Xu, Z., Lin, X. (2014). A reliable routing protocol based on QoS for VANET. In *The 7th IEEE/International Conference on Advanced Infocomm Technology*, Fuzhou, 21–28. doi: 10.1109/icait.2014.7019524.

4

Performance Evaluation of the Protocol for Collaborative Applications

4.1. Study background

In Chapter 3, a set of models necessary for the simulation of a routing protocol in a highway environment was presented. The selected models were used in the simulations and have formed the basis for the results that support the performance evaluation of the CBL-OLSR ad hoc routing protocol in this chapter.

The CBL scheme to be evaluated was designed for intelligent transportation systems (ITS). ITS applications are traditionally classified into three categories (Dar et al. 2010): comfort applications used by vehicle occupants, traffic optimization applications used by road operators to improve traffic conditions and finally safety applications intended to reduce or even eliminate the risk of accidents and their impacts. In this third category, which is of particular interest to us, Europe makes a distinction between primary road safety application services and secondary and tertiary road safety application services (ETSI 2013a, 2013b, 2018), as shown in Figure 4.1. Primary road safety application services include all of the applications involved in driver assistance. There are currently three of them. Two are activated based on the "Time To Collision" (TTC). These are the road hazard signaling (RHS) service (ETSI 2013a) and the collision risk warning (CRW) service, which concerns longitudinal CRW (LCRW) (ETSI 2013b) and intersection CRW (ICRW) (ETSI 2018). The third service concerns the onboard signaling of road signs in vehicles (*In-vehicle Signage*). Its time constraints are less critical. Secondary safety applications are triggered just before and after the accident in order to minimize impacts by reducing in particular the risk of further accidents. Finally, tertiary safety applications address emergency calling after the accident. With connected

Design and Evaluation of Ad Hoc Routing Protocol,
by Martine WAHL and Patrick SONDI.

autonomous vehicles, RHS and CRW services, in particular, and in addition to the information originating from embedded systems (Paret and Rebaine 2019), will take advantage of the information from adjacent vehicles by way of vehicle-to-vehicle (V2V), and more generally of vehicle-to-everything (V2X) links. The objective is to improve the accuracy of their own localization on the road (i.e. their "ego-localization") and the extent of their perception of their environment. Extended perception and ego-localization applications are two of the application scenarios that will be utilized for the system performance evaluation of CBL-OLSR in this chapter.

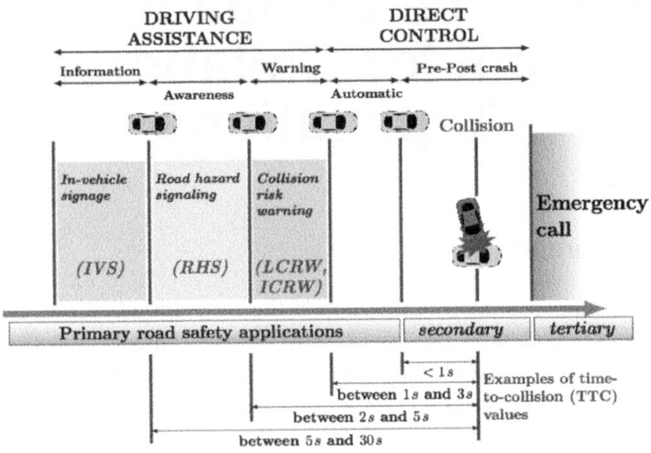

Figure 4.1. *Road safety applications and examples of TTC values according to ETSI (2013b). For a color version of this figure, see www.iste.co.uk/wahl/design.zip*

In order to facilitate the design of communication services for road safety applications in Europe, ETSI has started by defining two first types of messages for V2X applications, the messages from the *Cooperative Awareness* (CAM) basic service and those from the *Decentralized Environmental Notification* (DENM) basic service. CAM (ETSI 2019b) messages are periodically transmitted within a one-hop neighborhood. They have a transmission periodicity comprised between 100 and 1 000 ms depending on the use case. More specifically, they are used for broadcasting presence and position information from ITS stations to other ITS stations located in their direct communication vicinity (i.e. within "one hop"). The data transmitted in CAM messages may originate from the *Vehicle Data Provider* (VDP) or from the position and time manager and be collected upon receipt by the vehicle *local dynamic map* (LDM, Figure 4.4). DENM (ETSI 2019c) messages are transmitted after an event is detected. This event may be related to a road hazard, the driving environment, or traffic conditions. Sent from one ITS station to other ITS stations within an area of interest, they can be periodically repeated until the end of the event in a neighborhood that can comprise several hops. They serve as warnings to other road users of the occurrence of this event. Examples of events that trigger DENM messages include the "electronic brake light" or the "ice on the road" events. As communication services are specified and developed, other messages have since been progressively defined by ETSI in addition to CAM and DENM messages. For example, the report on the European project called *Transition Areas for Infrastructure-Assisted Driving* (Rondinome and Correa 2019) defines sets of V2X messages that are then reused by ETSI, the European standardization body. This is the case, for example, with CPM (*Collective Perception Message*) messages, whose performance evaluation can be found in the technical report (ETSI 2019a) on the analysis of the collective perception service. In our evaluation, we do not use CAM, DENM or CPM messages as defined by ETSI but UDP (*User Datagram Protocol*)/IP encapsulated messages.

Very early on, the scientific community has strived to explain the communication needs of V2X and I2V (*infrastructure to vehicle*) of different use cases. This is the case with Chen et al. (2010) and ETSI (2009), for example. In Table 4.1, we have mentioned applications listed by Chen et al. (2010) that require ad hoc V2V communications, periodically transmitted or following the occurrence of an event. These applications include two that present latency time constraints of 20 ms, namely, the application with periodic messages for driving in convoys (*platooning*) and the application for pre-accident detection, whose messages are transmitted when an event is received. Due to the low latencies required, the current ETSI definitions of DENM and CAM messages do not meet the time requirements of these applications. In the European *Mobile and wireless communications Enablers for the Twenty-twenty Information Society* (METIS) project, Fallgren and Timus (2013) recommended significantly more severe latency constraints for time-critical cooperative driving applications that focus on traffic efficiency and safety (*Test case 12: traffic efficiency and safety*), such as those of platooning and automated driving. The advertised end-to-end latency requirements are less than 5 ms for messages with a size of approximately 1,600 bytes for all V2X transmissions, with data transmitted either upon event or periodically at approximately 10 Hz (every 100 ms), and in a context where relative speeds between mobile objects can reach up to 500 km on the highway (Fallgren and Timus 2013). In particular, the following two data traffic models, recommended by ETSI and IEEE, were used in Europe for the performance study (ETSI 2012) of the CSMA, STDMA and MS-Aloha medium access algorithms for ad hoc communications in VANETs. These two models simulate a periodic CAM message transmission at a frequency of 2 Hz with 800-byte packets (ETSI configuration) and a periodic one at a 10 Hz frequency with 300-byte packets (IEEE configuration). This IEEE configuration of application traffic demand is reutilized in one of our application scenarios for the purposes of evaluating the performance of CBL-OLSR compared to other protocols.

Safety applications	Type	Transmitted information						Mode			
	V2I (and V2V)	Vehicule position	Heading, steering	Speed	Acceleration/deceleration	Indicator	Others	Periodic	Event	Latency (ms)	Range (m)
Intersection collision warning		✓	✓	✓		✓			✓	∼100	≤300
Cooperative collision avoidance		✓	✓	✓	✓			✓		∼100	≤150
Emergency braking		✓	✓	✓	✓				✓	∼100	≤300
Highway junction assistance		✓	✓	✓				✓		∼100	≤250
Blind spot detection		✓	✓	✓				✓		∼100	≤150
Pre-accident detection							✓		✓	∼20	≤50
Public transport priority at traffic lights		✓	✓	✓					✓	∼1,000	≤1,000
Platooning	✓	✓		✓			✓	✓		∼20	≤100
Adaptive speed control		✓						✓		∼100	≤150
Approaching emergency vehicle warning							✓		✓	∼1,000	≤1,000
Post-accident warning	✓						✓		✓	∼500	≤300

Table 4.1. *V2V communication requirements for road safety applications (Chen et al. 2010)*

This chapter therefore focuses on the performance evaluation of the CBL-OLSR protocol within the context of cooperative applications aimed at improving road safety. It presents several analyses of the CBL-OLSR communication system that were carried out. After a presentation in section 4.2 of the configuration of the mobile nodes used in the models, section 4.3 introduces a study of the structural and routing performance of CBL-OLSR. This study does not include application traffic: its objective is the validation of the implementation of the CBL scheme in an OLSR process in OPNET *Riverbed Modeler*. Then, the evaluations in section 4.4 take into account data traffic originating from the application layer (following the OSI layer ISO terminology). A first application, section 4.4.2, implements the model of an application data rate in accordance with the recommendations of IEEE (ETSI 2012): 300-byte messages are sent at a 10 Hz frequency. The purpose of the study is to achieve a comparative evaluation of the implementation of CBL-OLSR in OPNET *Riverbed Modeler* with those native to the AODV and OLSR protocols. In addition, an interested reader can find in Wahl et al. (2021) a comparison with the OPNET *Riverbed Modeler* GRP (*Geographic Routing Protocol*) protocol described in Zhiyuan (2009) and that of QOLSR (*Quality of Service OLSR*) defined by Sondi et al. (2014). The two evaluations in section 4.4 then focus on a collective perception application and an application combining data traffic for collective perception and another for ego-localization. The perception application will involve the broadcasting of one-hop perception data traffic, and the ego-localization application will combine two traffic flows, one in single-hop broadcasting and the other in multi-hop propagation in a single traffic direction via the CBL relay nodes.

4.2. Configuration of the mobile nodes

In all of the application scenarios presented in this chapter, the simulated systems are ad hoc networks of vehicles equipped with IEEE 802.11p cards in the 5.9 GHz frequency band. The transmission rate is 12 Mbit/s and the transmitter power and receiver sensitivity parameters, shown in Table 3.16, make it possible to obtain a communication distance of 500 m in order to correspond to the IEEE 802.11p specification for *Dedicated Short Range Communications* (DSRC). Depending on the applications studied and the scope of the performance evaluation, up to three node models are introduced: a CBL-OLSR node model, an OLSR node model and an AODV node model. All three implement the processes of the TCP/UDP/IP protocol stack and IEEE 802.11p radio technology implemented by OPNET *Riverbed Modeler*. Opting for the models provided by default in the network simulator ensures full result reproducibility. The CBL-OLSR process utilized is the result of our implementation of the CBL scheme within a copy of the native OLSR process of OPNET (section 3.6.4). The only changes made to the native OLSR and AODV processes concern the modification of the process initialization step of these protocols to allow their reset during simulation in order for them to support the ".trj" multi-trace trajectories explained in section 3.5.1.

Three distinct node models are used for comparing the performance of CBL-OLSR, OLSR and AODV protocols, which differ in the routing model implemented (Figure 3.29). OLSR and CBL-OLSR protocols are proactive protocols, and AODV is a reactive protocol. As such, OLSR and CBL-OLSR build and maintain paths in the background so that they can reach any node in the network when an application requires it (Chapter 2, p. 26 and 46). On the other hand, AODV only performs path searching when a request is made for a transfer of application data, and it remembers active paths only (Chapter 2, p. 27).

The routing mechanisms in OLSR and CBL-OLSR rely on the election and dynamic maintenance of relay nodes using two types of messages: local broadcast of HELLO messages for one-hop neighborhood discovery by all nodes, and propagation by relay nodes of control topology messages (TC messages) for updating the routing tables of the network relay nodes. In an OLSR network, a node selects as a multipoint relay node (MPR) only the smallest subset of its one-hop neighbors with which it maintains a symmetrical link and which enables it to reach any node within its two-hop neighborhood (Figure 2.5a). The information from these MPRs is then stored in its routing table. More precisely, the MPRs play two roles: one, they broadcast control traffic (TC message) in the network so that each MPR node is aware of paths to any node in the network at all times, and two, to relay application messages to their destination node(s).

CBL-OLSR was introduced in Chapter 2 (sections 2.5.2 and 2.5.3). The establishment of these links can be summarized as follows. At any given time, a CBL-OLSR node is in one of the following states: leaf versus branch. A branch node is a relay node and has a similar function to the OLSR MPR node in terms of application data transfer and TC messages. By default, a node is in the leaf state. In this state, it seeks to establish a leaf-branch connection with a branch node in its one-hop neighborhood. In the absence of such a node, it elects one from the leaf nodes within its one-hop neighborhood, thus triggering the change of state of that chosen node from leaf to branch node. The chosen node will truly become its branch after detecting the link symmetry. A node remains in the branch state as long as it remains the elected one of another node. In this state, it seeks to establish a branch–branch link in order to construct a chain of relays in the communication network, thus upgrading a leaf node to the branch state when there is no suitable branch node. The creation and maintenance of links between nodes follows path invariants, and more specifically its dynamics (see section 2.5.2.4).

OLSR and CBL-OLSR share the parameters of Table 4.2. Two parameter configurations were modeled. The parameterization B (case B) corresponds to the default values (Vd) in Clausen et al. (2003) in which a HELLO message is sent every 2 s. The parameterization A (case A) corresponds to half of the default values ($\frac{1}{2}$ Vd) for case B and therefore to a transmission frequency of every second for HELLO messages. The parameterizations A and B are used in section 4.3.1 during the validation step of the CBL-OLSR implementation in OPNET *Riverbed Modeler*. The parameterization A is then adopted from section 4.4 for evaluations with application traffic transmission.

An AODV node only performs a route search on request, that is, when it has data to transmit. Therefore, it first broadcasts a *route request packet*, (RREQ) within its one-hop neighborhood (TTL value initiated to one, Table 4.3). A RREQ receiver node knowing a path to the destination sends back a packet with a route response (*route reply packet* (RREP)); otherwise, and if the TTL counter value set by the requester enables it, it forwards in turn the RREQ request packet within its one-hop neighborhood. As long as it does not receive an RREP response packet and within the TTL threshold value, the requesting node will extend its search step by step to several hops per TTL value increment, each time by returning an RREQ request.

Attributes	Case B (Vd)	Case A ($\frac{1}{2}$Vd)
Interval between two HELLO messages (HELLO_INTERVAL)	2 s	1 s
Interval between two TC messages (TC_INTERVAL)	5 s	2.5 s
Neighbor hold time: Vtime (3×HELLO_INTERVAL)	6 s	3 s
Topology hold time (3×TC_INTERVAL)	15 s	7.5 s
Duplicate message hold time	30 s	15 s
Willingness	*Willingness Default*	
Addressing mode	IPv4	

Table 4.2. *OLSR and CBL-OLSR parameter values (Vd: default values indicated in RFC 3626)*

Attributes	Value	Attributes	Value
Route request retries	5	*TTL start*	1
Route request rate limit	10 packets/s	*TTL increment*	2
Active route timeout	3 s	*TTL threshold*	7
Hello interval	(1; 1.1 s)	*Local repair*	Yes
Packet queue size	*Infinite* | *5 packets*	*Hello loss*	2 s
Net diameter	35 hops	*Addressing mode*	IPv4
Node traversal time	0.04 s	*Timeout buffer*	2
Route error rate limit	10 packets/s		

Table 4.3. *AODV parameter values. Two AODV variants are modeled in section 4.4.2, one with an infinite queue size (Packet queue size) (denoted by AODV-Qinf), the other with a queue size of 5 packets (denoted by AODV-Q5)*

Periodic local broadcast of HELLO messages by the intermediate nodes of a recent route allows a path to be stored for some time (*active route timeout*, Table 4.3) to an IP destination in the routing table (this route is deleted if it has not been used or refreshed within the allotted time). If, after a certain number of route search attempts specified in the *route request retries* attribute of Table 4.3, no path to the IP destination is discovered by the source node, the latter deletes all packets stored for that destination from its queue. This deletion is performed regardless of the size of the AODV process queue that stores packets from the application to be transmitted for the time it takes to discover a path to their destination. Two AODV variants are modeled in section 4.4.2, one with an infinite queue size (*packet queue size*) (denoted AODV-Qinf) and the other with five packets (denoted AODV-Q5). Two variants are chosen so as to assist in the analysis of AODV protocol results, in comparison with the proactive OLSR and CBL-OLSR protocols. This value of five packets for AODV-Q5 was inspired by the study of Patel and Shah (2015) on the impact of queue size on the throughput and the packet deletions in an AODV network. Apart from this value of 5 for the *packet queue size* attribute, the parameter values reported in Table 4.3 are the default values in Perkins et al. (2003).

Networks		R1		R2		R3			R5		R4
Scenarios	1	2	3	4	5	7	8	13	12	11	10
Density	–	+	++	–	+	–	+	–	++	≈	≈
Average number of nodes	25	102	198	50	205	46	189	67	137	116	128
Ms1 (case B)	45.8	84.5	91.7	45.2	81.9	58.3	75.7	51.4	71.6	67	63.8
Ms1 (case A)	35.5	84	91.3	43.6	82.4	54.5	74.2	49.7	70.8	65.3	62.4
ΔMs1(Op–Ma)	–2.7	+2.5	+1.2	+9.9	+3.4	+17.4	+0.2		–		–
Ms2 (case B)	1.53	1.68	1.73	3.30	3.62	2.92	3.20	3.86	3.30	3.69	5.06
Ms2 (case A)	1.36	1.40	1.52	3.01	3.24	3.11	3.97	3.21	2.29	3.63	4.43
ΔMs2(Op–Ma)	–0.32	+0.07	+0.18	–0.45	+0.39	–1.77	–0.93		–		–
Ms3 (case B)	8.49	8.68	8.75	6.57	8.34	5.26	10.5	10.35	11.72	8.36	9.15
Ms3 (case A)	10.4	11.7	11.7	8.35	8.62	5.78	9.52	11.07	17.33	10.41	10.83
ΔMs3(Op–Ma)	+1	–2.4	–2.7	–1.25	–6.18	+0.08	–0.38		–		–
Ms4 (case B)	0.79	5.41	11	0.797	4.47	1.12	3.04	1.08	2.51	1.99	1.72
Ms4 (case A)	0.542	5.29	10.7	0.775	4.74	1.07	2.88	1.01	2.46	1.89	1.66
ΔMs4(Op–Ma)	–1.838	–0.88	–0.14	–1.515	–0.77	–1.05	–1.67		–		–
Ms5 (case B)	60.4	59.6	62.1	59.5	57.2	49	45.9	71.4	63.7	59.7	56.1
Ms5 (case A)	66	61	60.5	61.2	57.5	50.6	46.5	69.5	64.5	61.3	57.7
ΔMs5(Op–Ma)	+9.7	+7	+5.2	+3.6	+4.1	–1.7	+2.1		–		–
Ms6 (case B)	33.8	39	42.2	30.9	26.4	25.4	24.6	11.6	11.7	8.83	9.22
Ms6 (case A)	28	40.5	41.8	35.9	31.4	28	27.7	24.2	27.3	17.5	17.8
ΔMs6(Op-Ma)	–1.4	+1.1	–1.1	+5.9	–4.9	+4.4	–4.3		–		–

Table 4.4. *Average values of Ms structural performance metrics obtained between 150 and 500 s for CBL-OLSR in OPNET (Cases B and A, Table 4.2) and ΔMs_(Op-Ma) values for scenario A of their differences from those obtained in MATLAB (Rivoirard 2018)*

4.3. Validation of the CBL-OLSR implementation in OPNET *Riverbed Modeler*

The purpose of a structural evaluation of a protocol is to analyze what is related to its structure. Concerning the CBL-OLSR network, its composition in terms of leaf nodes, branch nodes and chains, as well as the robustness of its links should be analyzed. More specifically, the goal of section 4.3.1 is to verify that for the same road network and when there is no application traffic, the structure created by the CBL-OLSR network in OPNET is the same as the one created by the CBL scheme in MATLAB. This study will be based on the analysis of six structural metrics. The following study, section 4.3.2, focuses on a comparative evaluation of the routing performance of CBL-OLSR and CBL. This comparison will be carried based on seven routing metrics. Finally, the analysis of the CBL-OLSR implementation concludes, section 4.3.3, with the evaluation of the sensitivity of CBL-OLSR to the propagation model. This will be done based on the variations in the average values of the structural and routing metrics observed following the change in the propagation model.

4.3.1. *Structural verification*

A first implementation of the CBL protocol scheme carried out in MATLAB first led to validating the imagined concept (Rivoirard 2018; Rivoirard et al. 2018a; Wahl et al. 2021). In a second step, the use of the OPNET software *Riverbed Modeler* enabled the modeling all of the layers (according to the terminology of the ISO OSI reference model) of an ad hoc mobile communication station in order to evaluate the CBL scheme in the presence of application traffic. Once CBL has been implemented in OPNET in a copy of an OLSR process (CBL-OLSR process), the CBL scheme must be verified for structural compliance with its implementation in MATLAB.

Six structural metrics (Ms), characteristic of the CBL scheme, have been specified. These six structural metrics are:

– Ms1 (%), the percentage of leaf nodes in the network;

– Ms2, the average number of chains in the network;

– Ms3, the average number of branch nodes per chain;

– Ms4, the average number of leaf nodes per branch;

– Ms5 (s), the average duration of a node in the branch state;

– Ms6 (s), the average time in seconds that a leaf node remains attached to a branch node.

The values of these structural metrics obtained by simulation are listed in Table 4.4 for each of the A and B configurations of the OLSR parameters defined in Table 4.2.

Case B specifically corresponds to the nodes broadcasting a HELLO message every 2 s, as recommended by default in RFC 3626 of the OLSR V1 protocol (Clausen et al. 2003), and case A corresponds to broadcasting a HELLO message every second.

The ΔMs_(Op-Ma) values in Table 4.4 are the result of the difference between the value of the Ms metric obtained in OPNET for the CBL-OLSR network and the value of the same Ms metric measured in MATLAB for the CBL scheme. This comparison was made for the R1, R2 and R3 network scenarios (section 3.5, Chapter 3) in case A for a 1 Hz broadcast frequency of HELLO messages. It reveals little variation between the CBL implementation in OPNET and MATLAB. The few differences can be imputed to the way how the radio and queue management technology process the transmission and reception of routing messages. Therefore, we can approve the integration of the *clustering* CBL scheme into the OLSR process of OPNET *Riverbed Modeler*.

The Ms values in Table 4.4 are the result of the measurements of the structural metrics of CBL obtained for the CBL-OLSR network in OPNET *Riverbed Modeler* when (Table 4.2) each network node broadcasts a HELLO message every 2 s (case B column), on the one hand, and when this one-hop broadcast is performed every second (case A column), on the other hand. The average percentage of leaf nodes (Ms1) in the VANET network is linked to the percentage of branch nodes. The higher it is, the lower that of the branch nodes are (system relay nodes), and a lower proportion of relays reduces the communication resources consumed by the propagation of application traffic in the broadcast mode network (*broadcast*). The high values of Ms1 in medium- (S2, S5, S8) and high-density (S3, S12) scenarios reveal a percentage of branch nodes that does not exceed 30% of the total of nodes in the network. This reaches 38% for scenarios whose density varies over time (S10, S11), and finally, 65% for those with low densities (S1, S4, S7, S13) for which the nodes, sparse within the road networks, struggle to participate in the construction of a complete chain per direction. The number of branch nodes is independent of the road traffic density. It varies little with the density for the same road network, unlike that of leaf nodes (Figure 4.3): when a node becomes a branch, it tries to link to two other branch nodes, one upstream of its position and the other downstream, taking into account the direction of vehicle traffic and the communication range of the radio technology; nodes that do not participate in the construction of a chain are leaf nodes, either attached to a branch node that they have elected as such or isolated when no other nodes can be found within communication range. For the same reason, Ms1 also varies little with the periodicity of neighborhood discovery HELLO messages.

The average number of chains in the VANET network (Ms2 metric) depends on the modeled road network. Only one chain is expected in the R1 network with a single flow direction, two in the R2 and R5 (A27) networks with two flow directions, four in the R3 network which includes a section of motorway with two flow directions and two junctions with a single flow direction, and finally four in the R4 network with two flow directions in the A27 and in the RD90. However, Ms2 also varies according to the density and dispersion of road traffic (Tables 3.11 and 3.12). Its average values are higher than the target values for the R1, R2, R4 and R5 networks, but lower for the R3 network. All of these networks are prone to chain breaks caused by topological changes and resulting from the relative motion of vehicles (lane changes, overtaking, roadway entry or exit). In addition, for low-density scenarios (S1, S4, S7 and S13), the number of nodes is sometimes not enough to correctly create a chain in one direction. Scattered throughout the network, the nodes can be separated by distances greater than the radio communication distance, causing chain breaks. This scattering is further pronounced in the R3 network by the presence of an inbound and an outbound junction for both scenarios, which otherwise have little traffic. The deviation of the values of Ms2 from the target values increases when the transmission periodicity of HELLO messages is 2 s (Ms2 case B) due to a greater delay in the awareness and therefore in accounting for topological changes in the network: a lower periodicity (Ms2 case A) allows the nodes to respond more quickly to the neighborhood changes induced by node dynamics. With the exception of scenario S8 of the R3 network, an increase in the number of branch nodes per chain can be observed in the case of a 1-s HELLO message periodicity (Ms3 case A) compared to the values in case B. This is consistent with the reduction of the average number of chains (Ms2) whose length increases with faster management and anticipation of breaks. The average number of leaf nodes per branch (Ms4) decreases only marginally, as the number of leaf nodes in the network varies little between cases A and B; the same is true regarding the branch nodes in the network (see the slight variation in the percentage of leaf nodes, Ms1 metric, previously observed).

The average duration of a node in the branch state (Ms5 metric) varies only slightly upwards or downwards regardless of the periodicity chosen (case B or A). The Ms5 duration is shorter in the R3 network than in the R1, R2, R5 and R4 networks because the node state changes more often. In R3, a branch node inbound from the junction at the entry of the highway becomes a leaf when entering the highway section; conversely, a leaf node can be elected branch node by its new neighbors upon entering the highway exit junction. These changes of state in the R3 network during the journey decrease the average duration of Ms5.

With the exception of the low-density scenario S1 for which it decreases, the average linking time of a leaf node to the same branch node (Ms6) is increased when the frequency of transmission of HELLO messages is double-fold (Ms6 case A); this particularly holds for the R5 and R4 networks for which the road geometry (A27 and RD90) was modeled on a map (Figure 3.8) and the traffic modeled on the A27 from real traffic data from the Interdepartmental Directorate of Northern Roads

(DIR Nord, Figure 3.14). Therefore, a more frequent update of the local tables of neighbors results in a more precise selection of branch nodes vis-a-vis the dynamics of road networks (Ms2 metric) and a stabilization of leaf-to-branch links (Ms6) in mobility scenarios whose geometric characteristics, and therefore network topologies, approximate as close as possible to the ground truth (the roads of these networks are not modeled by straight lines).

In conclusion, the results obtained in Table 4.4, in comparison with a previous modeling in MATLAB or in comparison with the values obtained in OPNET *Riverbed Modeler* for different transmission periodicities of HELLO and TC messages (Table 4.2) make it possible to validate the modeling performed in OPNET. However, no conclusion can be drawn to whether there is improvement or degradation of structural performance for the scenarios tested as a result of the chosen configuration A or B. Nevertheless, for scenarios where the road configuration is closer to ground truth, a shorter time interval between two HELLO message transmissions (case A) seems to improve the robustness of leaf-to-branch links (Ms6).

4.3.2. *Analysis of routing traffic generated by CBL-OLSR versus OLSR*

The CBL-OLSR and OLSR networks were modeled in a similar manner. They particularly differ in the way they elect relay nodes (MPR versus branch), the format of HELLO messages, packets and routing tables that have been adapted for the implementation of the CBL scheme in CBL-OLSR (Chapter 2, section 2.5.3). The path-finding mechanics involved are nonetheless the same. The comparison of the routing metrics of the two networks is therefore relevant.

The following seven routing metrics, whose average measured values are plotted in Table 4.5, were calculated between 150 s and 500 s once the network is stabilized (i.e. when the average number of vehicles on the road network has stabilized, and the same applies to the average routing traffic amount in the communication network):

– The routing load, Mr1 (kbit/s), is the rate in kilobit per second of routing messages transmitted over the network. It comprises HELLO messages sent, TC messages sent and TC messages relayed (forwarded).

– The rate of HELLO messages, Mr2 (kbit/s), is the measure of the flow in kilobits per second of HELLO messages sent over the network.

– The TC message rate, Mr3 (kbit/s), is the measure of the flow in kilobits per second of all TC messages sent and relayed over the network.

– The ratio of HELLO messages, Mr4 (%), is the ratio expressed as a percentage of the throughput of HELLO messages to the routing load.

Network	R1			R2		R5			R4
Scenarios	1	2	3	4	5	13	12	11	10
Density	–	+	++	–	+	–	++	≈	≈
Average number of nodes	25	102	198	50	205	67	137	116	128
Mr1 (case B: pH.2s)	11.3	61.3	179	29.6	232	43.6	126	106	115
ΔMr1 (case B)	+9.2	+213.7	+911	+66.3	+2048	+130.4	+738	+488	+591
Mr1 (case A pH.1s)	26.8	126	358	58.4	463	85.3	249	219	235
ΔMr1 (case A)	+1.2	+335	+1322	+116.6	+3717	+234.7	+1271	+418	+603
Mr2 (case B)	5.36	47.5	158	17.7	195	27.8	92.3	70.1	78.6
ΔMr2 (case B)	–0.38	–1.6	–3	–0.8	+12	–0.8	–1.2	–0.6	–0.9
Mr2 (case A)	10.7	93.2	309	32.5	379	52.1	174	134	149
ΔMr2 (case A)	–0.84	–3.2	–6	–1.6	–10	–1.9	–4	–3	–4
Mr3 (case B)	5.91	13.7	20.5	11.8	37.2	15.9	34.2	36.1	36.3
ΔMr3 (case B)	+9.69	+215.3	+919.5	+67.1	+2032.8	+131.1	+738.8	+488.9	+592.7
Mr3 (case A)	16.1	33.2	48.9	25.9	83.2	33.2	74.9	85.3	86.5
ΔMr3 (case A)	+2.1	+337.8	+1331.1	+118.1	+3726.8	+236.8	+1275.1	+421.7	+606.5
Mr4 (case B)	52.3	78.6	88.3	61.4	84.3	64.4	73.7	67.4	69.7
ΔMr4 (case B)	-23.5	–60.1	–72.2	–42	–74.7	–46.3	–61.7	–54.4	–57.9
Mr4 (case A)	43.8	74.7	86.1	57.1	82.4	61.9	70.7	62.6	64.7
ΔMr4 (case A)	–4	–52.9	–65.7	–38	–73.1	–44.1	–57.3	–41.1	–46.6
Mr5 (case B)	47.7	21.4	11.7	38.6	15.7	35.6	26.3	32.6	30.3
ΔMr5 (case B)	+23.5	+60.1	+72.2	+42	+74.7	+46.3	+61.7	+54.4	+57.9
Mr5 (case A)	56.2	25.3	13.9	42.9	17.6	38.1	29.3	37.4	35.3
ΔMr5 (case A)	+4	+52.9	+65.7	+38	+73.1	+44.1	+57.3	+41.1	+46.6
Mr6 (case B)	83.4	87.5	87.6	86.4	91	86.8	90.2	90.6	89.3
ΔMr6 (case B)	+5.1	+10.5	+11.3	+10.1	+8.3	+10.1	+8.5	+7.9	+9.3
Mr6 (case A)	88.3	91.1	91	88.1	92.2	88.1	91.6	92.6	91.6
ΔMr6 (case A)	+0.5	+6.5	+7.6	+8.4	+7.1	+8.9	+7	+4.9	+6.3
Mr7 (case B)	0.112	0.147	0.2	0.12	0.185	0.125	0.147	0.135	0.137
ΔMr7 (case B)	–0.002	–0.027	–0.044	–0.005	+0.019	–0.007	–0.005	–0.002	0
Mr7 (case A)	0.111	0.145	0.196	0.119	0.181	0.124	0.144	0.132	0.135
ΔMr7 (case A)	+0.001	–0.017	–0.012	–0.003	+0.181	–0.002	+0.026	+0.004	+0.009

Table 4.5. *Average values of Mr1 to Mr7 routing metrics obtained for CBL-OLSR and deviation of OLSR routing metrics therefrom, computed between 150 and 500 s for cases B and A in Table 4.2 (Rivoirard 2018). The values of OLSR routing metrics are inferred by summation of Mri and ΔMri ($i \in [1, 7]$)*

– The ratio of TC messages, Mr5 (%), is the ratio expressed as a percentage of TC messages throughput to routing load.

– The TC_relayed/sent ratio, Mr6 (%), is the ratio expressed as a percentage of TC messages relayed on TC messages sent over the network.

– The WLAN delay, Mr7 (ms), as defined by the OPNET simulator, is the average delay measured in milliseconds at the WLAN level. It is the end-to-end delay of all packets received at the MAC layers of all nodes in the WLAN network and forwarded to the higher layers. This includes the access delay of the MAC layer of the sending node to the WLAN and the receipt of all individual packet fragments by the destination nodes.

The graphs in Figure 4.2 utilize the values of the metrics Mr1 to Mr3 from Table 4.5 and order them as a function of the average number of nodes in the scenario. They show the occupancy proportion of HELLO (Mr2) and TC (Mr3) messages in the total network load (Mr1). The graphs in Figure 4.3 plot the average number of leaf, branch and MPR nodes per scenario, as well as their minimum and maximum values. These graphs are complementary to those in Figure 4.2 from which the results can be understood.

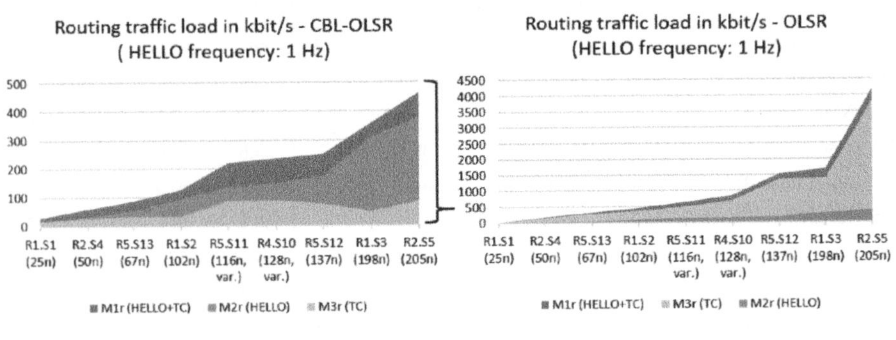

Figure 4.2. *CBL-OLSR and OLSR routing loads versus node density for scenarios R1, R2, R4, and R5 (case A, Table 4.5)*

Table 4.5 and Figure 4.2 show that the routing load increases with network density in the CBL-OLSR (Mr1 metric) and OLSR (respectively, "Mr1+ΔMr1") networks due to the increase in the size, on the one hand, and to the number of HELLO (Mr2 rate, respectively, "Mr2+ΔMr2") and TC messages (Mr3 rate, respectively, "Mr3+ΔMr3") with the number of nodes, on the other hand. First of all, the size of HELLO messages actually increases with the number of one-hop neighbors. If the throughput of HELLO messages (Mr2) is very slightly lower for OLSR, it is because of the 32 bits of additional information added to the CBL-OLSR

HELLO message for the position, node type (branch or leaf), speed, direction angle informations as well as the choice of nodes relative to the chain structure (address of the chosen branch node or the upstream node, Chapter 2, section 2.5.3.2). The traffic of the HELLO messages of the two networks is therefore similar modulo the size of the additional information. The size of the TC messages of a relay node (branch, respectively, MPR) increases with the number of nodes that have elected it; the CBL-OLSR Mr3 value thus follows the increase in the number of leaf nodes per branch node in the network (Ms1 metric, Table 4.5). In addition, the number of HELLO messages sent increases with the number of sending nodes or, for the same scenario, with their transmission frequency. Therefore, the Mr1 metric, just as the Mr2 metric, is almost twice as much for case A (1s HELLO_INTERVAL) compared to case B (2 s HELLO_INTERVAL) due to twice the HELLO message transmission frequency (2 × 0.5 Hz). On the other hand, the load of OLSR TC messages ranges from slightly higher for the S1 scenario to much higher values for all of the other scenarios compared to that obtained with CBL-OLSR networks. This load consists of the flow of messages emitted by the source nodes of these TCs and the flow of these messages relayed by the relay nodes.The more relay nodes there are, the greater the number of TC messages relayed for a TC message sent by a source node, and thus the greater the share of TC messages in the routing load (Figures 4.2 and 4.3). The proportion of TC messages relayed compared to those transmitted (Mr6) varies from about 88.3% to more than 92% for CBL-OLSR in case A and from 88.8% to more than 99% for OLSR, with the number of MPRs being higher than that of branch nodes as early as scenario S2 (it is similar only for scenario S1). For instance, for the S5 scenario, it is 168–201 MPR in the OLSR network compared to 29–45 branch nodes in the CBL-OLSR network. The results obtained therefore show that the lower routing CBL-OLSR load compared to that of OLSR results from the lower routing load in TC messages, in particular relayed ones. This result is due to a lower number of branch relay nodes in CBL-OLSR networks compared to the MPR relay nodes of the same OLSR networks, and this from a density of mobile nodes sufficient to cover the entire mobile network (in the context of our simulations, of at least 50 mobile nodes).

The Mr7 metric reveals that increasing the number of nodes slightly increases the WLAN transmission delay between the sending and receiving 802.11p interfaces. However, this increase remains relatively low, indicating that the communication medium is not saturated with the load of routing messages for any of the scenarios under study.

The study of routing traffic generated by CBL-OLSR and OLSR in OPNET *Riverbed Modeler* showed the decrease in TC message traffic by a factor of 3 and 17 depending on the scenario. This decrease is related to the small number of CBL relay nodes (branch) compared to the MPR nodes that generate these TC messages and relay traffic in *broadcast* mode.

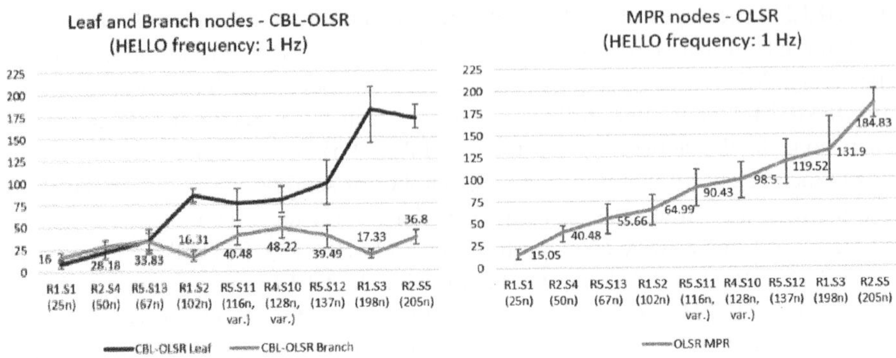

Figure 4.3. *Number of relay (CBL-OLSR branch, OLSR MPR respectively MPR) and leaf nodes (CBL-OLSR) versus node density for scenarios R1, R2, R4 and R5 (case A, Table 4.5)*

4.3.3. *Sensitivity to the propagation model*

The presence of neighboring vehicles can influence signal propagation. Therefore, to test the robustness of the CBL scheme, the simulations previously carried out with the low- to high-density scenarios S1 to S3 of the R1 road network were redone, this time with the implementation of the OLOS model of Abbas et al. (2015), as presented in section 3.6.3. The variation in the results obtained for the structural and routing metrics is reported in Table 4.6. With this OLOS model, a receiver node will successfully receive packets from a transmitter node located in a direct line and at a distance of no more than 500 m (the radio communication range), but will only receive packets from a transmitting vehicle at a distance of no more than 210 m when a neighboring vehicle sits between it and the transmitter within a 4-m wide rectangular area.

The largest metric variations are obtained for the low-density S1 scenario where nodes are further scattered in the network and far apart. This distance of the nodes causes few chain breaks (very small increase in Ms2) in a sparse network that struggles to create them, but a decrease in the average number of branch nodes per chain (δs3, Table 4.6) and a 9.1% increase in the number of leaves (δs1) along with a slight increase in the average number of leaves per branch (δs4). As matter of fact, the presence of vehicles obstructing the communication between two nodes more than 210 m apart will reduce the number of adjacent nodes to a node, and the impact induced on the structure will not be compensated due to the lack of high local density. Shorter chains are created with fewer branch nodes (decrease of Ms3), affecting their stability: the average time in the branch state decreases by 4.4 s (δs5). The decrease in the number of branch relays leads to a decrease in TC message traffic (δr3), and in particular of the ability to relay them (δr6), and therefore in the routing load (δr1).

	Structural metrics δsi=Msi(OLOS)–Msi(LOS)						Routing metrics δri=Mri(OLOS)–Mri(LOS)						
R1 network	δs1	δs2	δs3	δs4	δs5	δs6	δr1	δr2	δr3	δr4	δr5	δr6	δr7
S1	+9.1	+0.02	–0.58	0.251	–4.4	+6.9	–2.9	–0.1	–2.9	+5.5	–5.5	–1.7	+0.001
S2	–0.9	+0.05	+0.5	–0.31	–0.4	–4	0	+0.4	–0.7	+0.3	–0.3	–0.3	0
S3	–0.9	+0.23	–0.6	–1.08	–2.2	–6.2	–4	0	–3.8	+1.2	–1.2	–1.3	–0.001

Table 4.6. *δsi (respectively δri) results from the subtraction of the average values of the Msi structural performance (respectively routing Mri) metrics obtained for the CBL-OLSR network with an OLOS and LOS propagation model (R1 network scenarios; HELLO message periodicity of 1s – case A; computation performed between 150 and 500 s)*

Smaller but significant variations also appear with the high-density scenario S3. Due to the high density, it is not possible to keep the average number of branch nodes per chain, which slightly decreases (δs3), with the same order of magnitude as in the S1 case. It does not prevent the increase in the number of channels in the network (increase of Ms2). The number of leaves, in the network (δs1) and per branch node (δs4), is reduced in favor of an increase in relay nodes. The negative values of δs5 and δs6 indicate a change in the robustness of the link between a leaf node and its branch node with a decrease in its average duration of 6.2 s, but also in that of chains with nodes that remain 2.2 s less on average in this state. The robustness of the chain is nonetheless less affected than in the low-density S1 case. Despite the slight increase in the number of branch nodes in the network, the routing load decreases due to fewer TC messages being sent and retransmitted (δr1, δr3, δr5, and δr6). The decrease in the percentage of TC messages retransmitted out of those (δr6) sent suggests that the capacity of the network to relay information broadcast to all is reduced.

The medium-density S2 scenario is less impacted, with a slight increase in the number of branch nodes per chain (Ms3 metric) along with a decrease in the number of leaf nodes in the network. On the other hand, a significant reduction of four seconds in the average duration of links between a leaf node and its branch node is to be noted (Ms6 metric). However, the impact on the routing metrics of S2 is negligible, and zero for the total load on the network, because the chain structure is not or only slightly affected: the variation $\delta s2$ is negligible and the decrease in the duration in the branch state is relatively small.

In summary, taking into account obstruction phenomena to signal propagation due to the presence of at least one vehicle between a transmitter node and a receiver node shows with this implementation of the OLOS model a decrease in the stability of the leaf-branch links for each of the scenarios. It is also accompanied by a reduction in the stability of the chains created for the S1 and S3 scenarios, respectively, with low and high densities, which seems to disrupt the broadcasting of TC messages throughout the network. Finally, it should be noted that the average WLAN delay remains stable for each of the road scenarios S1 to S3, because in these networks, which are not saturated by the routing traffic load (section 4.3.2), this delay depends, when there is no additional application traffic, on the number of leaf and branch nodes of the scenario, the sum of which remains constant and the average number per type does not vary much.

4.4. CBL-OLSR system performance analysis with transfer of application data

4.4.1. *Introduction to collaborative perception and ego-localization applications*

Safety applications of ITS will require reliable services for vehicle positioning, environment perception and communications. Vehicular network applications currently rely primarily on infrastructure networks. However, due to the risk of infrastructure failure, the development of new communication services that can rely solely on V2V communications is necessary. In the perspective of full level-5 autonomous vehicles (*full automation*) according to the terminology used by the Society of Automotive Engineers (SAE, standard J3016) (Shuttleworth 2019; Paret and Rebaine 2019), that is, driverless vehicles and where humans are merely passengers in a vehicle and do not take part in the driving process, it will be compulsory to make sure that on the highway, for example, in the event of infrastructure failure, the driving mission of vehicles will proceed in degraded mode. This mode relies solely on their V2V communication links and their on-board sensors to ensure vehicle and passengers safety. At the very least, this objective would involve bringing vehicles to a halt, perhaps in rest areas, in order to free up the motorway lane until road traffic can be safely resumed. In the presence of an infrastructure failure, shutting down a driverless vehicle system could then require maintaining the accurate self-localization functions of vehicles in their environment and those of collective perception of the environment.

Collective perception aims to increase the local perception capabilities of vehicles beyond the limits of their own on-board perception sensors, whose maximum range is currently limited to about 100–200 m (Idrac 2018). Extending the perception of a vehicle to those of other vehicles requires making use of the ability of the vehicle to connect with other vehicles (V2V), or even other users and connected entities of the road (V2X), in particular the infrastructure (V2I, I2V), in order to then build a so-called collective or cooperative perception of the environment. Idrac (2018) stated that, for the time being, "automated vehicles get their bearings by means of combining a perception of the environment (horizontal signaling, restraint devices, vertical signaling) by native sensors (cameras, radars, lidars) and satellite tracking linked to on-board mapping". However, vehicles are not able to anticipate all kinds of situations such as those where the vehicle is engaged in an overtaking maneuver while an animal or a person, hidden by another vehicle, unexpectedly crosses a road. Therefore, a fully autonomous vehicle, namely without control and the ability to manually take over the driver at any time cannot be designed at the moment. A perception extended to the perception of other vehicles through the sharing of each of their perception of the environment would give the possibility to each connected vehicle to build and maintain a LDM (ETSI 2014a). This latter would integrate, after merging information, the elements perceived by the vehicles located within its communication range and those locally perceived through the sensors of the vehicle (Figure 4.4).

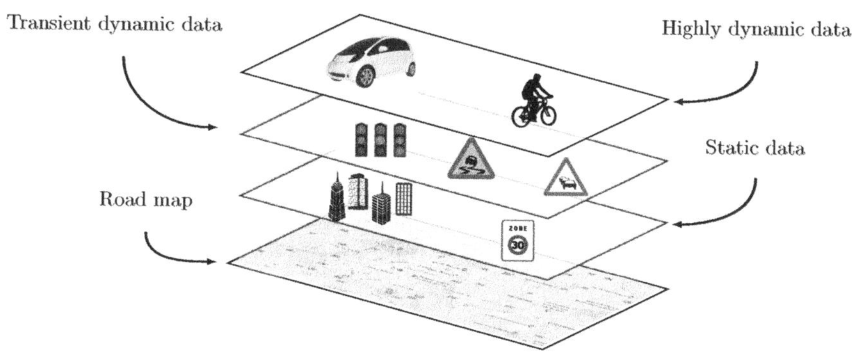

Figure 4.4. *Concept of extended map (local dynamic map (LDM)). Diagram inspired by Shimada et al. (2015). For a color version of this figure, see www.iste.co.uk/wahl/design.zip*

Two complementary concepts have been defined by ETSI with the aim of improving the vehicle's knowledge concerning its environment through the use of V2X links. The concept of "cooperative awareness" of our environment is the subject of the standard by ETSI (2014b). It refers to the broadcasting by road users and roadside infrastructure of their current state in terms of position, dynamics and characteristic. The concept of "collective" or "cooperative perception" (CP) has been more recently introduced by ETSI in a technical report (ETSI 2019a). It refers to the sharing with surrounding vehicles, by way of V2X technology, of information about the perception of a vehicle environment obtained by a system comprising environment perception sensors such as lidars, radars and cameras. ETSI (2019a) thus proposed the definition of cooperative perception messages (CPMs) for the transfer of information relating to objects located in the vehicle environment and detected by on-board sensors. When the evaluation of a collective perception application was undertaken in 2015, little information about these applications was available. For this reason, the cooperative perception study presented in this chapter will be based on data from Shimada et al. (2015) and Günther et al. (2015) and not from ETSI (2019a), which was published later.

Ego-localization refers to the capability of the vehicle to localize itself in its environment. The vehicle knowledge of its precise location is essential for the establishment of its local dynamic map, and therefore for the application of collective perception. The localization accuracy of the satellite-based global positioning system (GPS), which incurs an error of about 10 m (Kaplan and Hegarty 2005), may suffice for navigation assistance to the driver of a vehicle, but not for an autonomous driving system. In particular, it does not allow the GPS host vehicle to determine which lane it is on. Several studies proposed techniques that used different sensors embedded in the vehicle and an on-board mapping system. One such technique, called *cooperative map matching* (CMM) (Pollard and Gingras 2012), proposed the exchange of positioning information between neighboring vehicles through V2V communications. This technique was adopted by Rohani et al. (2016) who then used CMM to improve vehicle positioning. Rohani et al. (2016) proposed that this step be completed with a second approach called *dynamic base station differential GPS* (DDGPS) for which vehicles act as "dynamic base stations" in order to achieve a "differential GPS positioning". The choice of this ego-localization application will allow us to evaluate the capabilities of the CBL scheme to offer in parallel a local broadcast service and a data propagation service in a single flow direction.

The following sections focus on three application scenarios. The first concerns the periodic transmission of application data from multiple source nodes to multiple destination nodes over an interval of 100 s. The purpose of this study is to compare the CBL-OLSR, OLSR and AODV routing protocols, in terms of load, throughput, delay and data lost by deletions. This evaluation will be based on a set of application metrics already implemented in OPNET such as network load, network throughput, WLAN delay and number of hops. The second scenario concerns a collective perception application. Its purpose is to evaluate the frequency of environmental perception message exchanges that would be adequate for the constraints of a collective perception application to be met both in terms of end-to-end delay and correct packet delivery ratio. Finally, the third scenario focuses on the support concurrently provided by CBL-OLSR to several applications. For this scenario, we will add to the second application scenario the requirements of an ego-localization application. Again, the evaluations achieved will be based on the end-to-end delay and packet delivery ratio application metrics.

4.4.2. *Comparative analysis of CBL-OLSR with OLSR and AODV for* unicast mode *data transmission*

The objective of this first evaluation is to compare the performance of CBL-OLSR to that of two ad hoc routing protocols in the literature, AODV of Perkins et al. (2003) and OLSR of Clausen et al. (2003), in the presence of application traffic transmitted in *unicast* mode to several destination nodes.

The modeled application is that of a periodic transmission of application data in accordance with what could be the demand for an active safety application in a road environment. The size of the transmitted data follows the IEEE recommendations for communications in road environments (ETSI 2012): transmitting nodes transmit 300-byte data packets every 100 ms. Only 20 nodes, randomly selected once and for all during scenario modeling, thus before the simulation is initiated, are the sender or receiver nodes of the application data packets. But, all nodes in the network participate in the transmission of routing traffic and can relay application messages if they are identified as relays between the source node and the destination nodes. Application traffic was modeled in OPNET as follows: the transmission of application data between each source node and its 19 destination nodes is achieved by configuring a data flow of 300 bytes every 100 ms for each source-destination node pair.

4.4.2.1. *Simulation configuration*

The study was carried out in OPNET with the following simulation parameters. The mobility scenario is the medium-density S2 (Table 3.11) of the R1 road network in one traffic direction (section 3.5). This scenario is executed over a duration of 100 s, which corresponds to a distance traveled of about 3.6 km for a car going at 130 km/h. Over this distance and when the road network has reached the desired density, nearly 70 mobile nodes (400 trucks/h at 110 km/h and 2,000 cars/h at 130 km/h (Table 3.11 and Figure 3.15)) circulate on average simultaneously (compared to 102 on average over 5 km (Table 4.4)). Given that the simulation time of the road scenario is short, 70 nodes suffice for this simulation for which a set of 70 single-trace trajectories in the conventional ".trj" format of OPNET *Riverbed Modeler* has been created (Figure 3.17).

A simulation consists of a set of three successive simulations, each initiated with the same *seed* value used by the random number generator in OPNET *Riverbed*

Modeler. One of the three simulations executes CBL-OLSR as the routing protocol, the second OLSR and the third AODV. Each simulation is run for a scenario execution time of 100 s. A set of three simulations is run 10 times, with a separate value of *seed* for each new run (Figures 4.5(A)–(C)). A preset *seed* value allows us to set the first element of a sequence of numbers for which each number is computed based on its predecessor by the random number generator (actually "pseudo-random" numbers). The purpose of using a new *seed* each time a simulation is launched is to avoid being in a particular configuration that would give preference to one of the evaluated protocols. Each result curve relative to the metrics measured during the simulation of Figures 4.6 to 4.12B thus reports the average values calculated from the results of the 10 simulations.

The introduction of our trajectories based on SUMO traces means that vehicles gradually enter the road network. The expected vehicle density is then only reached after about 200 s (Table 4.5). To avoid this gradual increasing load of the road network to the planned density, the track of vehicles corresponding to this period has been truncated from the trajectory set used for this application: the positions of the vehicles at the initialization time $t = 0\ s$ of a simulation therefore correspond to those of date 200 s of the road network. The transmission of application traffic by the 20 nodes is triggered at the 30th second of the OPNET simulation for a period of 30 s; the first 30 s give the communication network time to stabilize. The application traffic activation period is thus included between times "30 s" and "60 s" of the result figures.

4.4.2.2. *Analysis of simulation results*

Application data are transmitted over the WLAN network after traveling through the UDP/IP stack. Traffic reaching the UDP level combines, for OLSR and CBL-OLSR proactive protocols (Figures 4.5(A) and (B)), data traffic transmitted by the application and routing traffic from the *manet_rte_mgr* process, which is a parent process of the OLSR and CBL-OLSR processes (Figure 3.29). Regarding the AODV protocol, traffic arriving at the UDP level only concerns application data (Figure 4.5C), path search is only enabled at the IP level by the AODV process, invoked by the *manet_mgr* process following the arrival of UDP/IP encapsulated data at the *IP_dispatch* parent process level (Figure 3.29). Figures 4.5(A)–(C) show that the sending of 19 $\times$ 300 bytes every 100 ms of application traffic by a source node correspond at UDP level to the encapsulation of 19 $\times$ 10 UDP packets per second, giving a total of 190 packets per second to be transmitted at IP level. These application data are either directly forwarded to the WLAN lower layers if there is a route in the AODV routing table, or is queued in the AODV queue if there is space available when a route search is activated and awaiting its result, or deleted if the AODV queue is already full. Modeling of two queue sizes was performed for AODV, one of infinite size for the AODV variant called AODV-Qinf, and one of five packet size for UDP/IP encapsulated application data (IP datagram) of the variant called AODV-Q5.

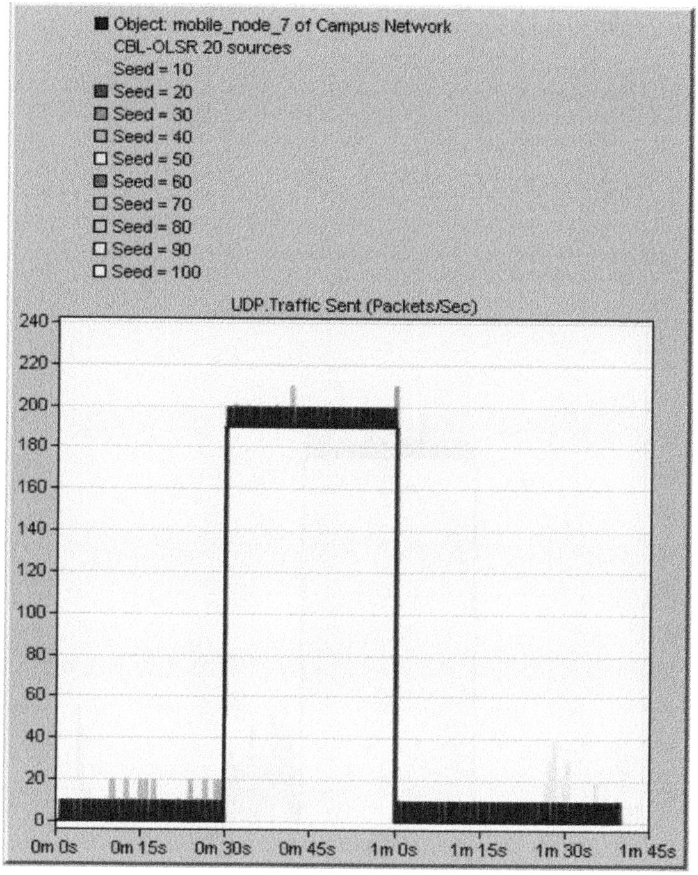

a) CBL-OLSR

Figure 4.5A. *Number of packets per second sent by the UDP transport protocol to the lower IP layer from source node 7 (results for each seed). For a color version of this figure, see www.iste.co.uk/wahl/design.zip*

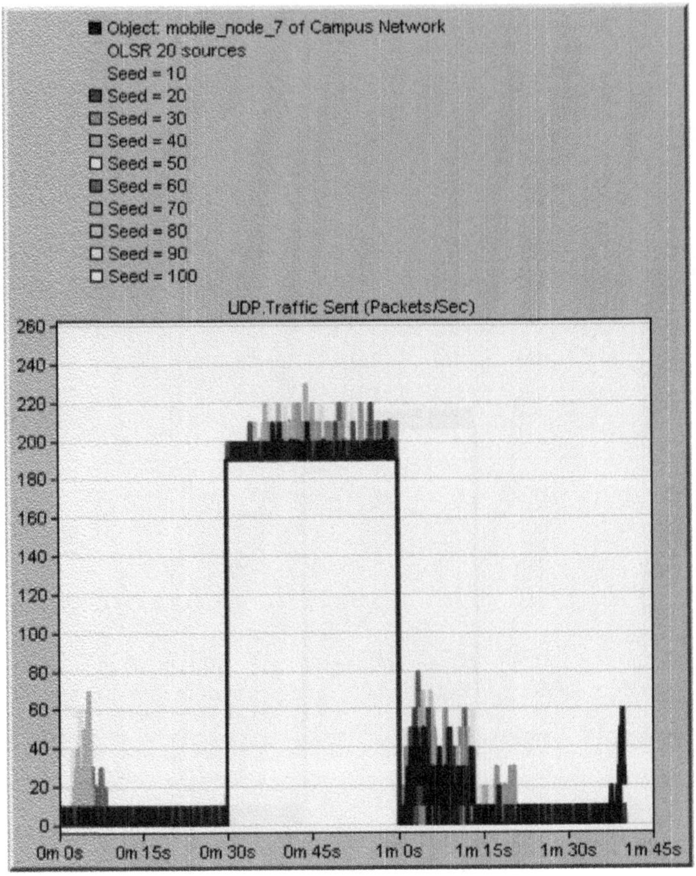

b) OLSR

Figure 4.5B. *Number of packets per second sent by the transport UDP protocol to the lower IP layer from source node 7 (results for each seed). For a color version of this figure, see www.iste.co.uk/wahl/design.zip (continued)*

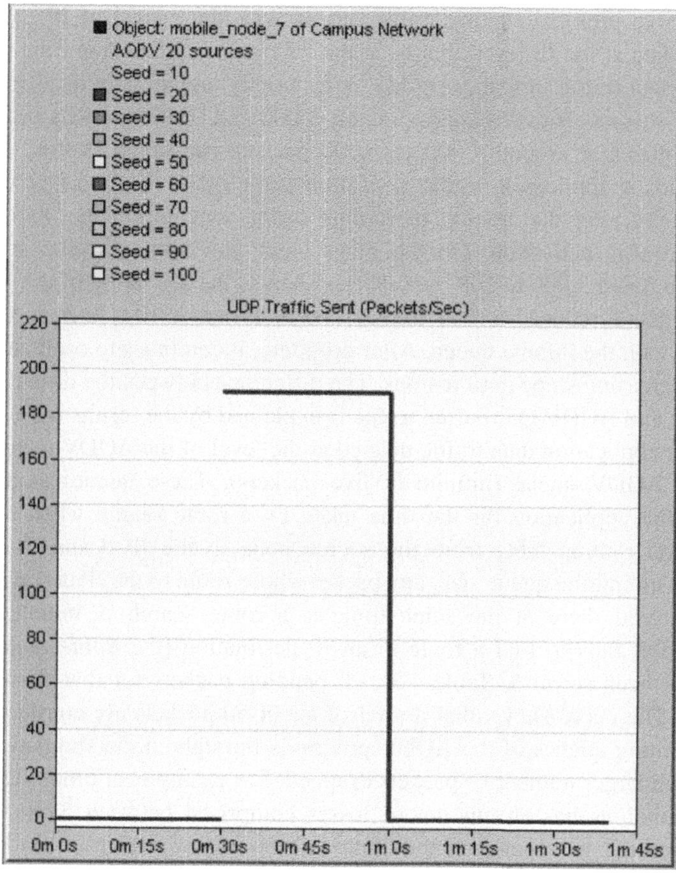

c) AODV

Figure 4.5C. *Number of packets per second sent by the UDP transport protocol to the IP lower layer from source node 7 (results for each seed). For a color version of this figure, see www.iste.co.uk/wahl/design.zip (continued)*

Figure 4.6 presents the average values of the "routing traffic sent" metric (Figure 4.6(a)), obtained by the CBL-OLSR, OLSR and AODV protocols, and the number of relay nodes of the proactive protocols (Figure 4.6(b)). In Figure 4.6(a), the AODV reactive protocol routing traffic begins with the arrival of IP encapsulated application data at the IP level, that is, at the 30 s request activation time to transmit application data traffic in *unicast* mode. After briefly and sharply oscillating during the first few seconds between approximately 1,000 and 10,000 packets per second, it quickly stabilizes at around 4,000 to 5,000 packets per second before decreasing from the end of application traffic activation (time 60 s). The size of the AODV queue does not alter the amount of routing traffic sent during the 30 s period of application traffic activation. On the other hand, the routing traffic sends drops sharply from about 3,000 to 100 packets/s as soon as the activation of the application stops for a queue reduced to five packets, but only about 10 s later and in similar proportions with the infinite queue. After dropping, it continues to oscillate for about 20 s on a decreasing slope until it stops. The difference between the drop times of the AODV-Qinf and AODV-Q5 routing traffic is explained by the results in Figure 4.7(b) representing application data traffic deleted at the level of the AODV queue for each size of the AODV queue (infinite or five packets). These queues store the data packets of the application for the time taken by a route search when no route is known in the routing table from the source node to the IP destination of those packets. For an infinite queue size, any packet whose route to the IP destination is not known is stored there at the same time as a route search is initiated. If after five attempts it fails to find a route to an IP destination (the *Route request retries* attribute has been set to 5, Table 4.3), all pending packets for that destination are eliminated. The curve shows that a total of about 50 packets are constantly purged from the waiting queues of the AODV processes throughout the simulation, with at times a much larger number of packets dropped. For example, at time 50.8 s each of the 10 simulations has an amount of losses comprised between 5,681 and 8,432 packets, resulting in a peak on the curve greater than 7,400 packets dropped on average. When the application activation stops (time 60 s), AODV-Qinf still has many packets to transmit in the network, unlike the AODV-Q5 network for which there are almost no data left to transmit with a local queue in nodes limited to five data packets. This difference in behavior between the AODV-Inf and AODV-Q5 curves after time 60 s can also be seen by repercussions in the curves of the measurements carried out at the WLAN level (Figures 4.9(b) and 4.11).

Routing traffic sent in OLSR through CBL-OLSR networks fluctuates throughout the simulation while remaining below 2,000 and 200 packets per second for OLSR and CBL-OLSR, respectively (Figure 4.6(a)). The continuous presence of routing traffic is due to the periodic transmission of routing messages necessary to constantly update the routes of these two proactive protocols. The traffic curve of CBL-OLSR oscillates very regularly between 100 and 200 packets per second, while that of

OLSR shows a higher amount of routing traffic with a curve that irregularly varies and contains a trough at the activation time of the application and a peak successively at the end of the activation. If we first disregard the abrupt fluctuations of the OLSR curve over a period of about 10 s after the activation time of the data traffic on the one hand and the time of the end of its activation on the other hand, we observe that, over this interval of 30–60 s, the curves of the two proactive protocols account for a slight decrease in routing traffic. This decrease can be explained due to the priority given to the transmission of application traffic in *unicast* mode over the retransmission of TC topology messages in *broadcast* mode. *Broadcast* mode packets are transmitted with jitter to avoid collisions, and will not be retransmitted over the 802.11p WLAN network in the event of a collision, unlike those sent in *unicast* mode. Parallel observations of this curve of the evolution of the number of MPR nodes during the simulation (Figure 4.6(b)) shows similar oscillations, in addition to a non-optimization of the number of MPRs during the application activation period. This is actually because, outside of the network activation and stabilization period, the optimal number of MPR nodes is about 40 nodes. Following the activation of application traffic, a short and abrupt drop of about 10 relay nodes is observed (Figure 4.6b), followed by an increase which brings the number of nodes to about 50–55 between times 35 and 65 s. However, the periodic transmission and reception of OLSR HELLO messages makes each node become aware of its one- and two-hop neighborhood. And the periodic transmission of topology control (TC) messages broadcasting the list of their neighbors that have chosen them as MPR enables each node to construct their topology table and deduce their routing table that will enable them to route packets to any destination (Naimi and Jacquet 2003). The non-optimization of the number of relay nodes during the application activation period indicates a loss of routing traffic at the WLAN level by collision, and more specifically of TC traffic due to the increase in the probability of collision on the TC messages whenever they are relayed by an MPR (Laouti et al. 2002). It is only after the end of this activation period and in about 10 s that the number of nodes returns to its optimal value, after the TC messages re-flow correctly, and in about 20 s that the routing traffic flow stabilizes. Regarding the routing traffic curve of the CBL-OLSR protocol (Figure 4.6(a)), it remains relatively constant despite a very slight decrease between 30 and 60 s, potentially due to a collision loss at the WLAN level of TC traffic in particular; the same mechanisms as those used in OLSR by the MPRs for relaying messages within the network through 1-hop broadcasts are implemented in CBL-OLSR at the level of the branch nodes. It should be noted that this slight decrease in traffic is not reflected in the average number of branch nodes whose selection by branch or leaf nodes is achieved, within a one-hop neighborhood, on the basis of node mobility criteria on a roadway: on average, their number remains constant in order to cover the entire R1 road section.

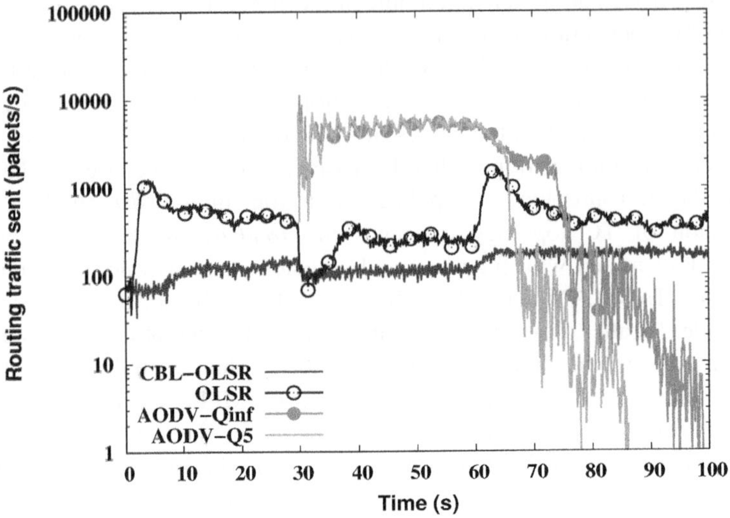

a) Routing traffic sent (base-10 logarithmic y-scale)

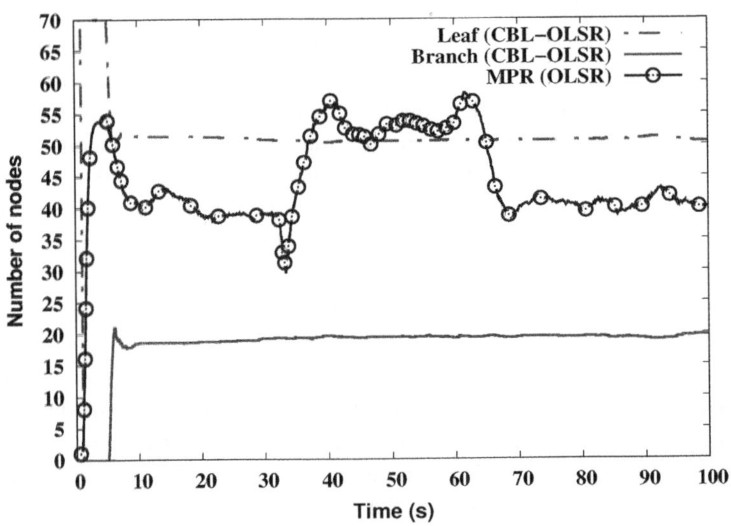

b) MPR nodes versus branch and leaf nodes

Figure 4.6. *Average values of routing traffic sent by CBL-OLSR, OLSR and AODV protocols and average values of the number of CBL-OLSR and OLSR relay nodes. For a color version of this figure, see www.iste.co.uk/wahl/design.zip*

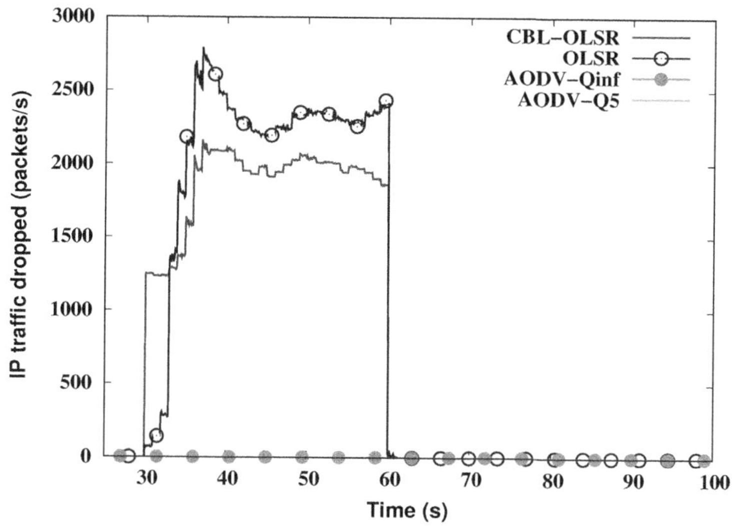

a) Traffic dropped at the IP level

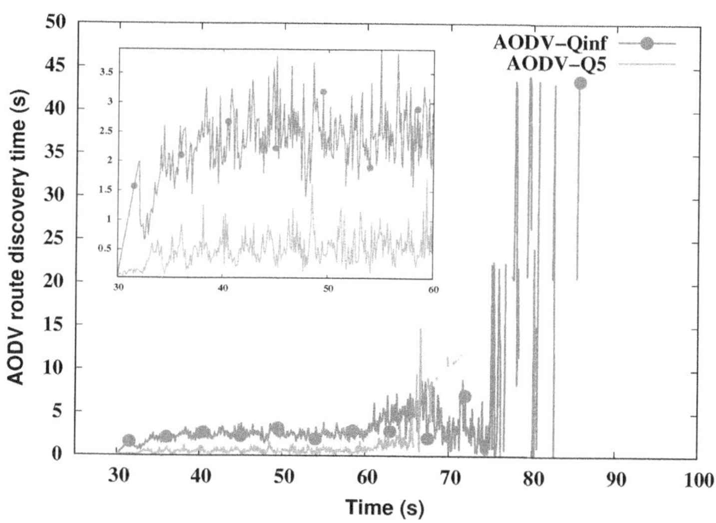

b) Traffic dropped at the AODV level

Figure 4.7. *Average values of dropped traffic at the IP level (in packets/sec) and at the AODV queue level (in numbers of packets) (20 sources). For a color version of this figure, see www.iste.co.uk/wahl/design.zip*

The curves in Figure 4.7 represent the amount of IP datagrams (IP data packets) deleted. The amount eliminated by all nodes in the OLSR and CBL-OLSR networks at the IP level is expressed as the number of packets per second (Figure 4.7(a)). The number dropped by AODV network nodes at the AODV queue is expressed as the total number of packets (Figure 4.7(b)). An IP datagram of the OLSR and CBL-OLSR protocols carries routing traffic or application traffic. For modeling these two protocols, IP datagrams can be deleted at the IP level when the maximum number of hops is exceeded, when an MPR (respectively, branch) node does not have a route registered in its routing table to the destination of the IP datagram, or when a non-MPR (respectively, leaf) node finds no relay (MPR, respectively, branch) node in its neighborhood while the destination of the IP datagram is more than one hop away. Figure 4.7(a) shows that CBL-OLSR loses slightly fewer packets per second than OLSR at the IP datagram level. This figure does not depict any deletion of IP datagrams for the AODV protocol, because in the AODV network, all application data are processed at the level of the AODV process where the IP datagrams of the application data are managed as well as the route searches necessary to forward them to their destination when they are not known prior to their transmission (Figure 4.7(b)). Figure 4.7(b) shows that AODV drops more or fewer data packets at the level of its queue depending on the size of its queue. Application packets are stored in the AODV queue waiting for an RREP response with a route to their destination following the transmission of an RREQ request (section 4.2). They are deleted when no routes are found within the allotted time and when the queue size is exceeded. With a storage memory of five packets per node, about 200 packets are regularly dropped at the AODV level in the AODV-Q5 network. These correspond to both packets for which there is no known route at the time of their request to be sent in the AODV routing table but which cannot be queued due to lack of space, and to pending packets for which the number of route search attempts has exceeded the allowed number of five attempts (Table 4.3). With an infinitely large queue, only about 50 packets are regularly dropped, with however a queue purge of more than 7,400 packets on average at time 50.8 s, as reported above in the study on routing traffic. According to Figure 4.8, the time to discover an AODV route from the arrival of a data packet to be transmitted to an unknown destination in the AODV routing table is in average less than one second for AODV-Q5, but around 2.5 s plus or minus 1 s for AODV-Qinf. This time corresponds to the average time spent by IP datagrams in the AODV queue before they are transmitted. It should be noted that, at this level, it is not possible to estimate whether one or the other of the protocols drops more application data packets because the IP datagram loss metrics in Figures 4.7(a) and (b) do not refer to the same set of vehicle traffic (application traffic only for AODV versus application traffic plus routing on the OLSR and CBL-OLSR side) and moreover they have a different unit of measurement.

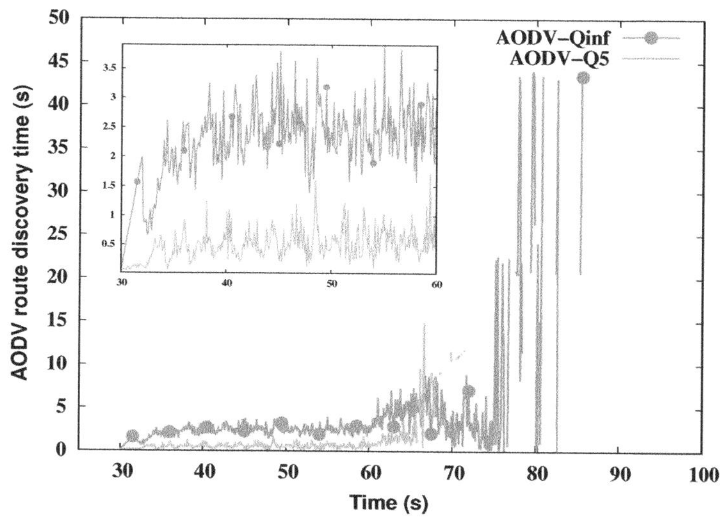

Figure 4.8. *AODV route discovery time (in seconds). For a color version of this figure, see www.iste.co.uk/wahl/design.zip*

The curves in Figure 4.9 illustrate WLAN load and throughput in bits per second. WLAN load (Figure 4.9(a)) is representative of the total load of the bits submitted at the WLAN layers (layers 1 and 2 according to the ISO OSI reference model) by the top-level layers of all WLAN nodes in the network. This load includes application and routing traffic for each of the networks under study, CBL-OLSR, OLSR and AODV. WLAN throughput (Figure 4.9(b)) corresponds to the total rate of traffic forwarded (pulled up) from the WLAN layers to the higher layers, across all WLAN nodes in the network. There are few differences between the load and throughput OLSR, CBL-OLSR and AODV-Qinf curves during the application activation period. We find that the AODV-Q5 load is lower than that of AODV-Qinf, due to higher data loss at the AODV-Q5 queue level, and consequently a lower amount of traffic related to route finding. However, the AODV-Q5 and AODV-Qinf WLAN rates are similar and correspond to the maximum bandwidth capacity of the WLAN network. Moreover, AODV networks are the ones that have the largest amount of data deleted at the 802.11p level, either due to exceeding the capacity of the queue or to exceeding the threshold of possible transmission attempts. The CBL-OLSR load is higher than that of OLSR and is similar to that of AODV-Q5 and lower than that of AODV-Q5 if we look at the total load presented at all WLAN nodes in the networks. On the other hand, the CBL-OLSR load is larger than that of other networks if this time we consider the metric corresponding to the load presented at the level of the WLAN technology by a single WLAN node (Figure 4.10).

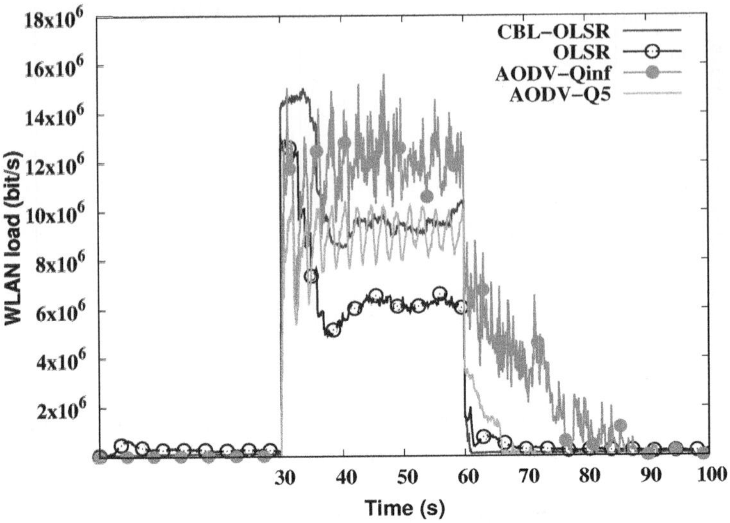

a) WLAN load

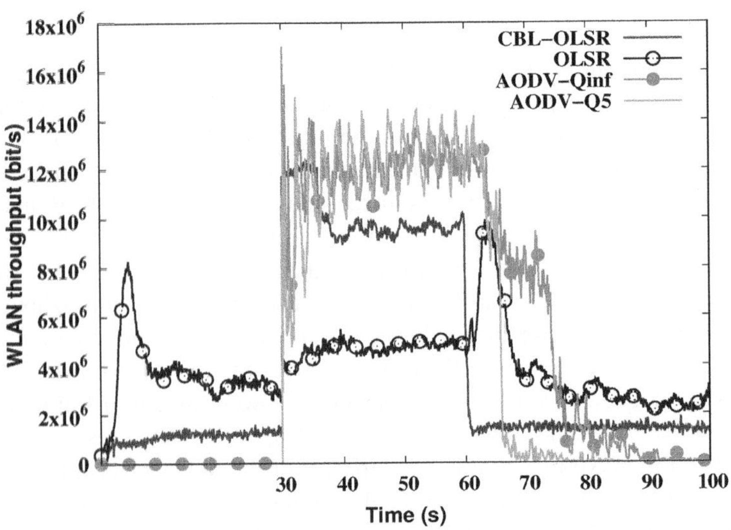

b) WLAN throughput

Figure 4.9A. *Average load, average throughput and deleted data at all WLAN nodes in bit/s (20 sources). For a color version of this figure, see www.iste.co.uk/wahl/design.zip*

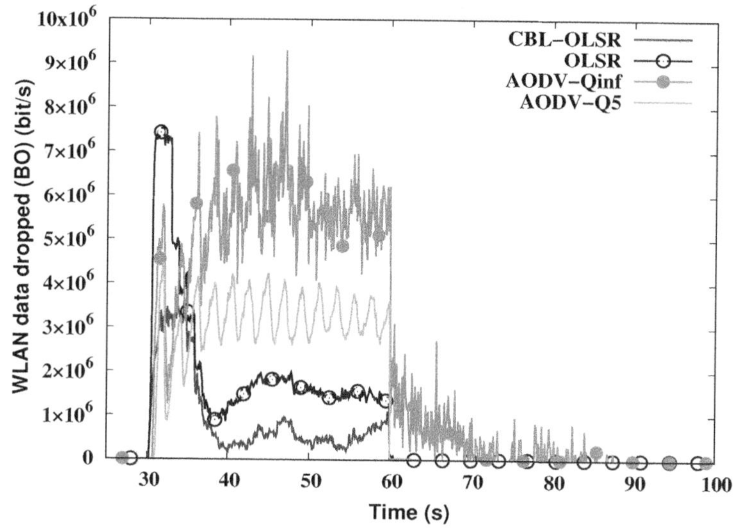

c) WLAN data deleted due to buffer memory overflow (BO)

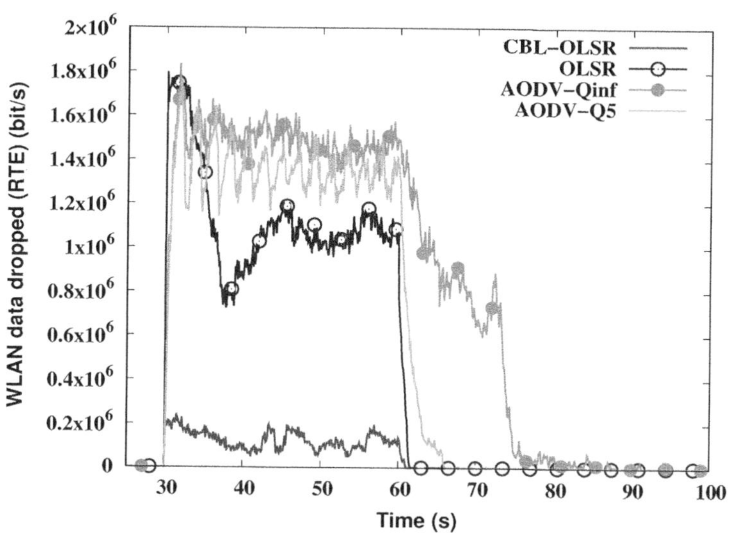

d) WLAN data deleted due to exceeding the threshold of retransmission attempts (RTE, Retransmission Threshold Exceeded)

Figure 4.9B. *Average load, average bit rate and deleted data at all WLAN nodes in bit/s (20 sources). For a color version of this figure, see www.iste.co.uk/wahl/design.zip (continued)*

This second metric, proposed by OPNET, is sized so as to measure the network load separately for each BSS (*Basic Service Set*). The WLAN throughput of CBL-OLSR is higher than that of OLSR, but lower than that of AODV (Figure 4.9). The WLAN throughput of CBL-OLSR stabilizes on average around 10 Mbit/s between times 40 and 50 s after reaching maximum bandwidth capacity during approximately the first 10 s of application traffic activation. This throughput is similar to the load presented at the WLAN level by all nodes, which is approximately 2 Mbit/s lower in the first few seconds when the transmission demand exceeds the capacity of the medium. The OLSR throughput is approximately 1–1.5 Mbit/s smaller than its load, which between times 40 and 50 s approximately corresponds to the amount of data of about 1 and 1.2 Mbit/s deleted when exceeding the transmit attempt threshold. In terms of data deletion at the WLAN level, CBL-OLSR is the protocol that suffers the least from losses either due to deletion for exceeding the threshold of retransmission attempts or due to buffer overflow at the WLAN level (Figures 4.9(c) and (d)).

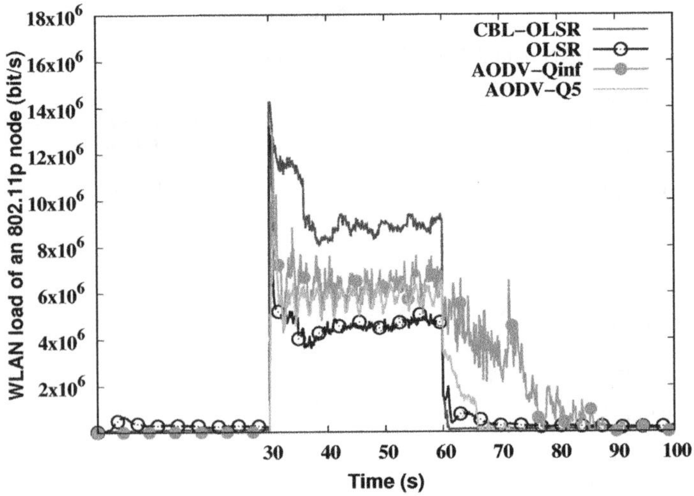

Figure 4.10. *Total traffic load in bit/s received by a WLAN BSS from layers above the MAC layer accepted and queued for transmission. For a color version of this figure, see www.iste.co.uk/wahl/design.zip*

Moreover, CBL-OLSR experiences the lowest number of retransmission attempts (Figure 4.11(a)) despite a pulse-shaped increase at the very end of the activation period. Its WLAN end-to-end delay is also the lowest of the three protocols; this is an important property for real-time application traffic flowing through a VANET network (Figure 4.11). To this delay, the AODV protocol adds the time it takes to discover a

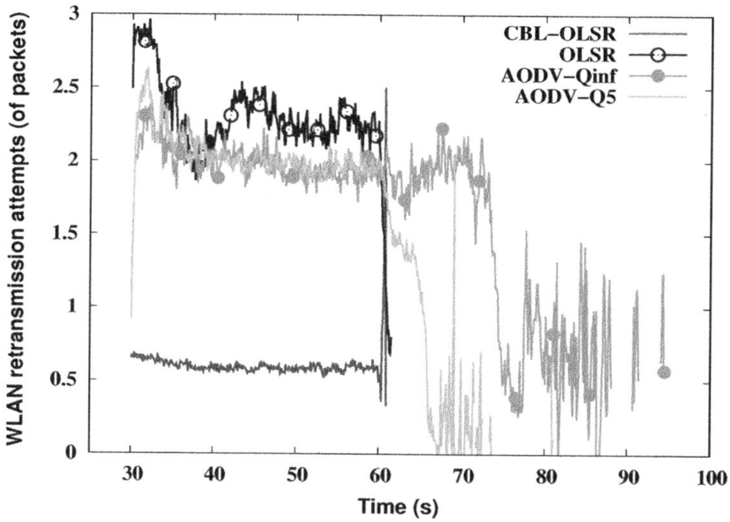

a) Total number of retransmission attempts in the network at the WLAN level

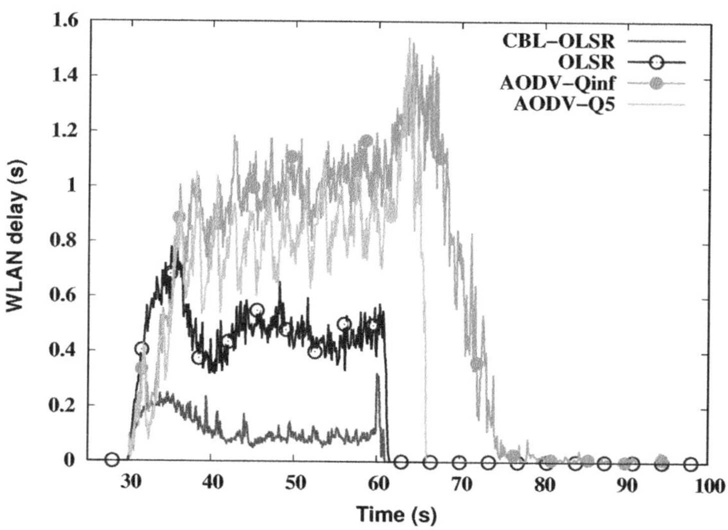

b) WLAN delay

Figure 4.11. *WLAN retransmission attempts (in numbers) and WLAN delays (in seconds). For a color version of this figure, see www.iste.co.uk/wahl/design.zip*

route to the packet destination when it does not exist in its routing table (Figure 4.8). The very high mobility of traffic on a highway configuration penalizes the AODV protocol in this respect. The path search time, of around 0.5 s for AODV-Q5 and from 2 to over 2.5 s for AODV-Qinf, increases the risk of obsolescence both for the route at the time of transmission, and for data and more specifically real-time data.

Another interesting metric is the number of IP hops relative to data packets reaching their destination. Figure 4.12(a) shows that it remains stable on average for the three protocols during the application traffic activation period, with an increase after this period for the two AODV networks. However, the high number of packet losses at the IP level (Figure 4.7) and the low number of hops (around two for CBL and AODV, slightly lower for OLSR) may suggest that part of packet losses are due to losses at the level of transfers to their destinations through the relays. This seems to be confirmed when comparing the average results obtained according to the number of transmitting sources: 10, 15 or 20 sources (Figures 4.12(a)–(c)). We observe that the average number of CBL-OLSR IP hops during the application activation period is greater than three and exhibits fluctuations between three and four hops with 10 sources (Figure 4.12(a)). It is greater than two hops, with fluctuations between 2 and 2.5 hops on average for 15 sources (Figure 4.12(b)); finally, it averages around two hops, with lower amplitude fluctuations for 20 sources (Figure 4.12(c)). These curves also show transmission delays that increase for AODV with the number of transmitting sources after the end of the activation period, but also a number of hops that can exceed 12. Consequently, all networks suffer from overloading the bandwidth available with WLAN technology.

In conclusion, simulations were performed for comparing the CBL-OLSR network with the proactive OLSR and reactive AODV networks using the same 802.11p WLAN technology, under the same simulation and node mobility conditions (highway traffic, one times three lanes in the same direction). The results show that AODV routing traffic is the most heavily impacted and present the highest delay. The OLSR proactive routing protocol shows better optimization of the routing traffic. However, the performance of CBL-OLSR is even better. Moreover, CBL-OLSR delivers the best performance on the delay metric. Data losses, on the other hand, at the CBL-OLSR IP level in particular, remain quite high, although slightly lower than those of OLSR. It may be useful to investigate whether introducing quality of service (QoS) criteria during branch node selection can reduce these losses while maintaining, or even improving, the balance between network load and throughput.

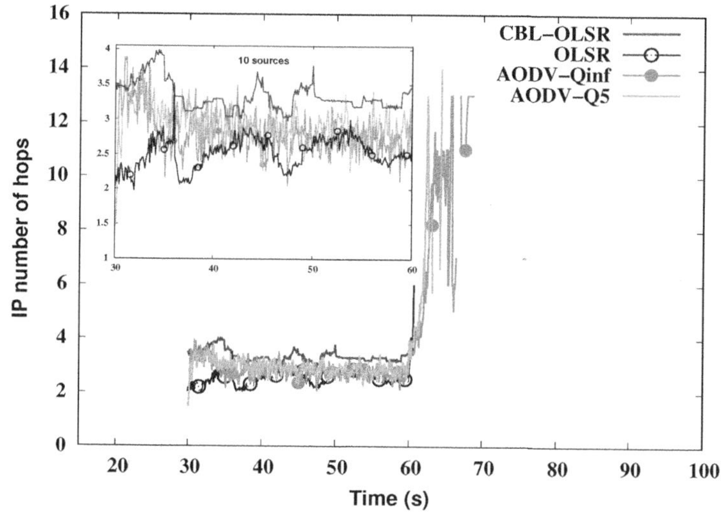

a) IP number of hops for 10 sources

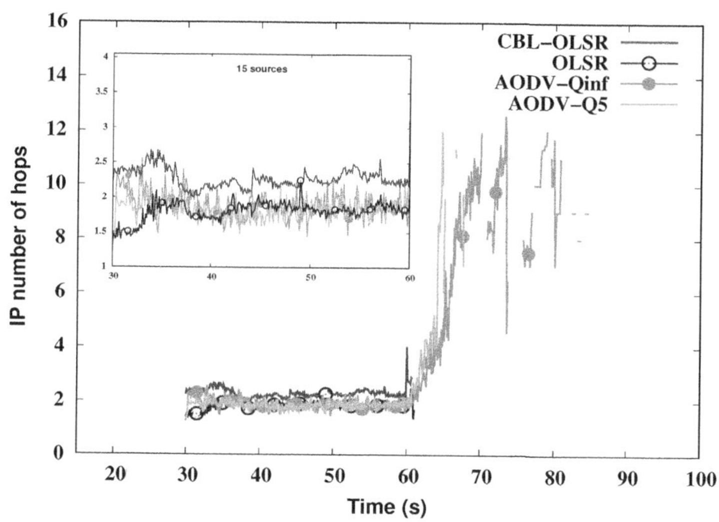

b) IP number of hops for 15 sources

Figure 4.12A. *Average number of hops at the IP level for 10, 15 and 20 transmission sources. For a color version of this figure, see www.iste.co.uk/wahl/design.zip*

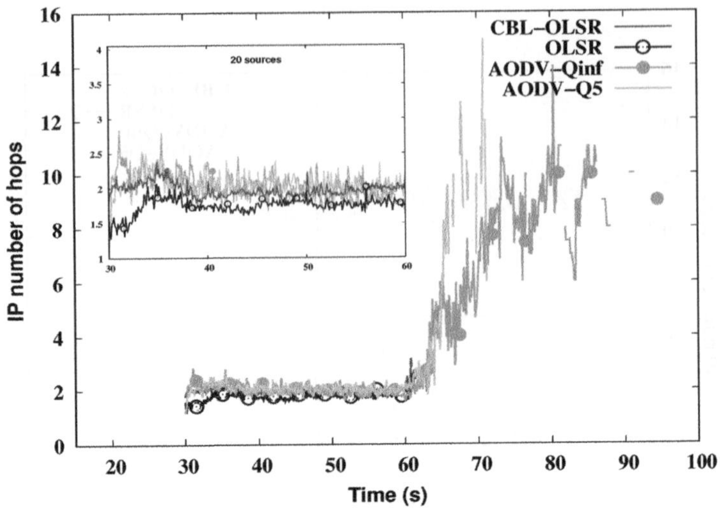

c) IP number of hops for 20 sources

Figure 4.12B. *Average number of hops at the IP level for 10, 15 and 20 transmission sources. For a color version of this figure, see www.iste.co.uk/wahl/design.zip (continued)*

4.4.3. *Performance evaluation of an extended perception application*

Collective perception has been the subject of various studies. Günther et al. (2015) and Günther (2018) have examined the problem concerning whether cooperative perception according to the penetration rate of communicating vehicles in the car market has interest. They show, with Figure 4.13, that periodic sending of HELLO or CAM messages suffices for the elaboration of a dynamic local map integrating all vehicles within the transmission range area of 300 m under consideration (the environmental awareness ratio is 1, the penetration rate is 100%). For a lower penetration rate, the use of several technologies will lead to reducing detection losses of the neighborhood and the contribution of the extended perception of any object becomes significant with less than 80% of vehicles equipped. For example, with a penetration rate of 40%, radar alone will only allow the vehicle to detect 20% of its neighborhood (the awareness ratio is at 0.2); beside receiving EPM (*environmental perception message*), it will detect 60%, and it would have detected only 45% with receiving periodic CAM or HELLO messages alone. Combining CAM messages and collective perception messages (EPM or CPM) will be particularly interesting to increase the reliability of local collective perception systems, in particular by providing a diversity of sources and information redundancy.

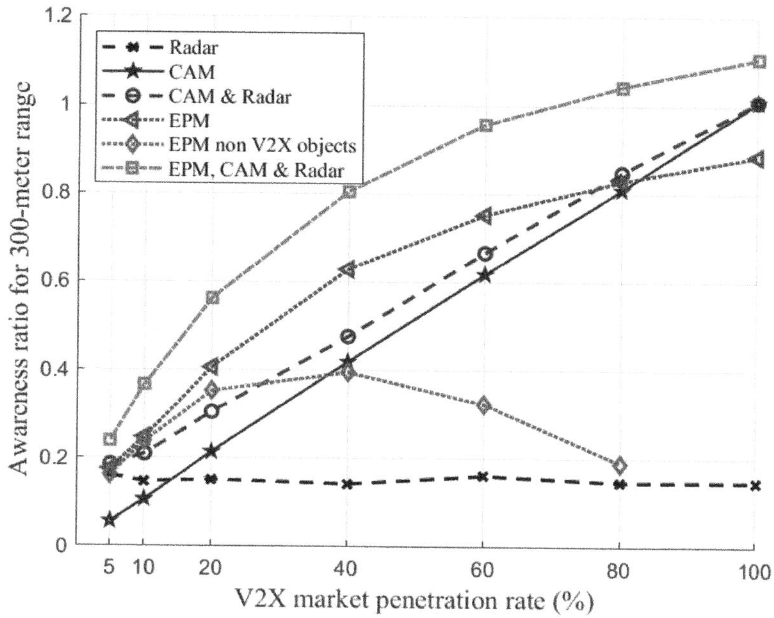

Figure 4.13. *Proportion of the vehicle neighborhood detected within a 300-m area versus the market penetration rate in terms of vehicles equipped with a 80-m range radar and with a maximum aperture of 60° and/or means of communication. Diagram based on Günther et al. (2015)*

In order to be useful, the additional information transmitted via an EPM must reach each vehicle within an end-to-end delay and with a correct packet delivery ratio (PDR) for the targeted application.

Consequently, the objective of this evaluation is to estimate the appropriate frequency of EPM exchanges for a collective perception application that satisfies both a low end-to-end delay constraint as well as a correct packet delivery ratio constraint for a given traffic scenario when only V2V means of communication are available.

The requirements for the collective perception application in terms of communication intervals (or transmission periodicity), throughput and delays are not formally established in the literature. Günther et al. (2015) suggested that the application transmitted EPMs within a 1-s time interval; their size should range between 37 and 709 bytes. Giang (2014) chose a frequency of 10 Hz (100 ms period) for each vehicle to send the following data: 20 bytes encoding position, speed and their statistical deviations; 22 bytes encoding the respective identification, position, velocity and variance values relating to the detected surrounding obstacles; 9 bytes encoding data relating to the characteristics, type and confidence indications of the road infrastructure; 4 bytes to encode the type of time, visibility and adequate safety distance; 8 bytes encoding traffic sign element data (position, type and nature of information); finally, 3 bytes for adding risk-level related information, risk information and operating procedures. Due to the large amount of information to be transmitted, the work of Giang (2014) established, through experimental results, that it was not possible to scale up the application for a vehicle density greater than 100 vehicles per kilometer. Finally, Chaabouni et al. (2016) proposed using 24 bytes for representing dynamic objects (detection time, position, velocity, acceleration and size) and 18 bytes for static objects (detection time, position and size), as described in Figure 4.4. The latter are the object sizes that were chosen for modeling the perception application in the CBL-OLSR network.

4.4.3.1. *Perception application model in CBL-OLSR VANETs*

For this study, the implementation of the application follows the approach proposed in Chaabouni et al. (2016). The collective perception application is carried out as follows with the CBL-OLSR network:

– Each CBL-OLSR leaf node periodically transmits an EPM to its branch node that contains data about static and dynamic objects detected in its environment.

– Each branch node, after merging the information received from its leaf nodes with its local information, returns the resulting data to its leaf nodes and branch nodes located just upstream in the same flow direction (i.e. the one following it). Furthermore, each vehicle has access to traffic information for the same driving direction.

In order to guide the direction taken by the transmitted messages, the *message type* in the CBL-OLSR message header complies with the format described in Table 2.2.

Modeling the size of packets transmitted by a leaf node and a branch node can be described as follows.

– The number of mobile nodes N_{mob} perceived by a CBL-OLSR leaf node is computed as the number of neighbor nodes within a 100 m radius around its position among the neighbors it detects via its communication interface. The number of static nodes N_{stat} detected by a CBL-OLSR leaf node is determined by a random variable comprised between 0 and 10. The size of the message transmitted by this leaf node is then given by: $N_{mob} * 24\ bytes + N_{stat} * 18\ bytes$, where 24 bytes and 18 bytes are the sizes of the moving and static objects, respectively, proposed in Chaabouni et al. (2016).

– For the computation of the size of the packets transmitted by a branch node, that is, after merging those received from its leaf nodes, each branch node collecting information from its leaf nodes and merging them before retransmitting them, it was assumed that traffic generated by the branch node is twice the size of the largest volume of traffic from the one generated by its leaf nodes.

4.4.3.2. *Simulation configuration*

The study was carried out in OPNET. The mobility scenario is that of high-density S12 (Table 3.12) of the R5 road network corresponding to a section of the A27 highway (Figure 3.8). All nodes are equipped with an 802.11p interface (Table 3.16) and execute the CBL-OLSR protocol (Rivoirard et al. 2018a; Wahl et al. 2021) (section 3.6.4).

Each simulation scenario lasts 300 s. Statistics are only collected from 200 s onwards, once the network is stabilized. EPMs are transmitted with a periodicity ranging from 50 to 500 ms (namely a frequency of 2–20 Hz) depending on the simulations. Ten different seed values are used for the random number generator (and therefore one simulation run per seed for each distinct periodicity), such that the sequences of numbers drawn do not favor a particular situation during the OPNET simulations.

Three application metrics were chosen to evaluate the performance of CBL-OLSR in the presence of the collective perception application. The first is the PDR, which represents the ratio of the number of packets successfully received to the total number of packets expected (equation [4.1]). This metric makes it possible to select from all transmission periods those that offer an acceptable PDR according to the requirements of the application domain:

$$PDR = \frac{\text{Number of packets successfully received}}{\text{Cumulative sum of packets sent}} \qquad [4.1]$$

The second metric is the IP end-to-end delay (IP delay). It corresponds to the difference between the time the packet was transmitted and the time it arrived at its destination. The appropriate transmission rate could be chosen to achieve an end-to-end delay that is consistent with the time requirements of the application. Its choice could also be modulated according to the speed of mobility observed in road traffic. The third metric explained earlier in section 4.3.2 is the WLAN end-to-end delay (WLAN delay).

The results collected were reported as a function of the periodicity of the EPMs in Figures 4.14, 4.15 and 4.16.

4.4.3.3. *Analysis of simulation results*

The results show that the best PDR of 91% is obtained when each vehicle transmits an EPM every 350 ms (Figure 4.14). The corresponding IP end-to-end delay is then approximately 88 ms (Figure 4.15). A 350-ms periodicity may however be too long for time-critical collective perception applications. Nonetheless, a periodicity of less than about 175 ms leads to IP end-to-end delays larger than the transmission periodicity. The end-to-end delay observed at the WLAN level (Figure 4.16) suggests that real-time EPM transmission every 50 ms can be achieved within a short time of 0.5 ms, provided however that the application operates directly at the level of the data link layer (layer 2), namely directly at the level of the 802.11p on-board radio interface, and that it tolerates a PDR of 88%.

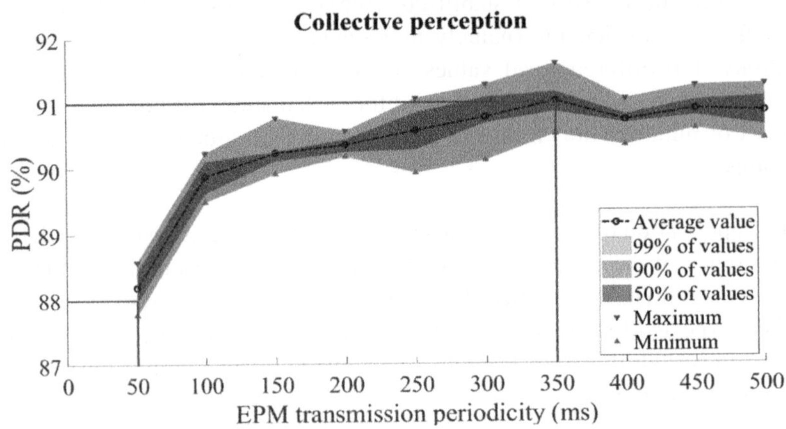

Figure 4.14. *Packet delivery ratio (PDR) for different EPM transmission periodicity. For a color version of this figure, see www.iste.co.uk/wahl/design.zip*

In conclusion, the study on the performance of V2V communications for a collective perception application shows that the application can operate in real time with very short delays (less than 0.5 ms) and a high packet delivery ratio of at least 88%, provided that it is implemented directly above the radio technology, that is, by bypassing the network and transport layers.

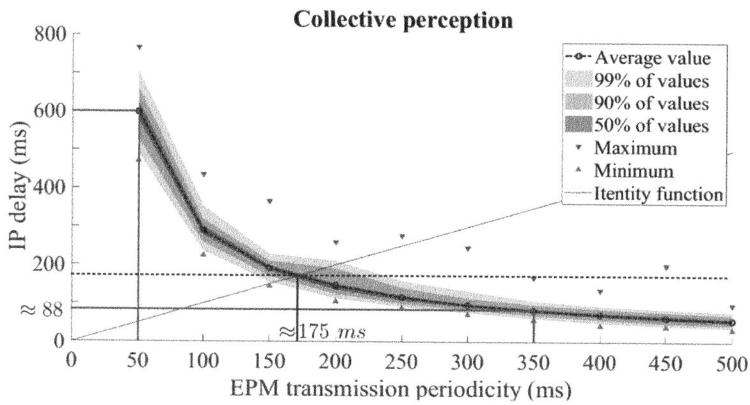

Figure 4.15. *IP end-to-end delay for different EPM transmission periodicity (in milliseconds). For a color version of this figure, see www.iste.co.uk/wahl/design.zip*

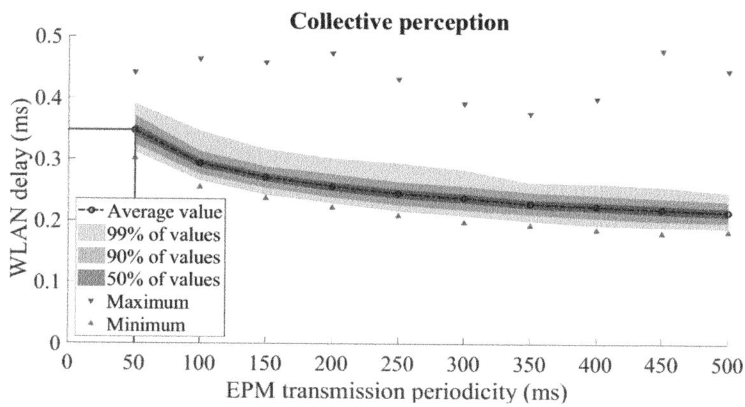

Figure 4.16. *WLAN end-to-end delay for different EPM transmission periodicity (in milliseconds). For a color version of this figure, see www.iste.co.uk/wahl/design.zip*

4.4.4. Multiple applications: extended perception and collaborative distributed localization

A high-performance localization system would benefit collective perception applications. Similarly, many studies on routing protocols, especially geographical ones, assume that a global localization service is in place informing each vehicle of the location of all other vehicles (Ashok et al. 2011; Sheu et al. 2012; Al-Ezaly et al. 2021). Such a service is based on an existing infrastructure and is therefore not accessible when only V2V communications are available. In this case, the information of each vehicle about the localization of the other vehicles is based solely on the exchange of their GPS location via V2V communication. A number of different systems strive to reduce the accuracy error of GPS below 10 m. For example, differential GPS (DGPS) improves the positioning accuracy of GPS using a differential method that uses a fixed known position as a ground reference. This approach requires however the deployment of an expensive infrastructure. Based on this approach, Rohani et al. (2016) proposed to use mobile, rather than fixed, reference stations that would make corrections to the pseudo-distance values received by the GPS and share them with one another in order to compensate for the common error of a GPS satellite according to an algorithm called DDGPS. This DDGPS cooperative algorithm would potentially be run in combination with another, namely the CMM algorithm. In this approach, the CMM algorithm, locally executed in any vehicle, aims to integrate into the mapping correlation algorithm of a vehicle the constraints of the road infrastructure imposed on collaborative vehicles in its same neighborhood within one communication hop, in order to improve the accuracy of its positioning by removing aberration areas. Section 4.4.4.1 is dedicated to defining the communication resource requirements for CMM and DDGPS application functions.

4.4.4.1. *Communication resource requirements of CMM and DDGPS applications of the method of Rohani et al. (2016)*

The purpose of *the CMM algorithm* is to refine the position of the vehicle calculated from the pseudo-distance measurements of the GPS receiver and to reduce their uncertainty by taking advantage of (a) the mapping constraints that apply to any road vehicle in use, (b) the fact that vehicles share via their V2V communication their pseudo-distance measurements within their one-hop neighborhood and finally (c) the following property, "the common errors $\varsigma_j^{(i)}$ of the pseudo-range measurements $\rho_j^{(i)}$ of each visible satellite j lead to the same bias in the GPS position for each vehicle (i) in the same neighborhood" (Rohani et al. 2016). Pseudo-distance measurements can be written as follows (equation [4.2]):

$$\rho_j^{(i)} = D_j^{(i)} - c\delta t^{(i)} + \varsigma_j^{(i)} + \eta^{(i)} \tag{4.2}$$

where $D_j^{(i)}$ is the actual distance between satellite j and vehicle (i), c is the electromagnetic wave speed, $\delta t^{(i)}$ is the clock synchronization error and $\eta^{(i)}$ is the uncommon errors.

First, to refine the value of the estimated position, the CMM algorithm of a vehicle (i) evaluates the GPS position of (i) ($\tilde{X}^{(i)} = [\tilde{x}^{(i)}, \tilde{y}^{(i)}, \tilde{z}^{(i)}]$) and the covariance of this position from the pseudo-distance measurement $\rho_j^{(i)}$ inferred from each signal received from the satellite j. In a second step, the CMM algorithm will reduce the uncertainty on the GPS position by comparing the new estimated position with the constraints of the road on which the vehicle (i) is traveling. A vehicle in motion has a very low probability of traveling outside the road infrastructure (roadside, median strip, etc.) and cannot travel on the site of a building, for example. By matching local road constraints to the estimated position, the CMM algorithm reduces the uncertainty of the GPS position. In a third step, the vehicle (i) computes the GPS position and the covariance on the position of a vehicle (v) in its neighborhood from which it receives the pseudo-distance information $\tilde{X}^{(v)}$, this information having been previously broadcast by (v) to its surroundings. By translating the road map constraints that apply to the estimated position of (v) to its own estimated position, the vehicle (i) further reduces the uncertainty about its own position. According to Rohani et al. (2016), the map constraints of any vehicle (v) can be copied to (i) when all vehicle positions are improved by "using the same group of satellites". The estimated position $\hat{X}^{(i)}$ is thus iteratively and constantly refined by transferring the constraints applying to vehicles in the neighborhood to the position uncertainty of (i). For the CMM algorithm to work, Rohani et al. (2016) relied on the assumption that any vehicle (i) that performs the CMM function broadcasts its own pseudo-distance measurements $\rho_j^{(i)}$ to its vicinity for each visible satellite j.

The DDGPS algorithm strives to compensate for the common error of each visible satellite j by locally computing information for correcting this error and sharing it with other vehicles (Rohani et al. 2016). First, each vehicle locally computes a pseudo-distance correction information $\Delta\rho_j^{(i)}$ for each visible satellite j using equation [4.3], and then transfers it to its neighborhood. In a second step, each vehicle (r) receiving a pseudo-distance correction information $\Delta\rho_j^{(i)}$ can correct its measured pseudo-distance values using equation [4.4] to create for each visible satellite j a set of corrected pseudo-distance values $\rho_{j,corr}^{(i)}$:

$$\Delta\rho_j^{(i)} = \hat{D}_j^{(i)} - \rho_j^{(i)} \quad [4.3]$$

$$\rho_{j,corr}^{(i)} = \rho_j^{(r)} + \Delta\rho_j^{(i)} \quad [4.4]$$

where $\Delta\rho_j^{(i)}$ is the pseudo-distance correction information of the j satellite computed by (i); $\hat{D}^{(i)}$ is the geometric distance between the position of the satellite j and the estimate of the position $\hat{X}^{(i)}$ by the vehicle (i); $\hat{X}^{(i)}$ the position estimate of vehicle (i) which may have been improved using the CMM algorithm; $\rho_{j,corr}^{(i)}$ a pseudo-distance corrected value of the visible satellite j generated by vehicle (r).

A vehicle (r) can receive pseudo-distance correction information $\Delta\rho_j^{(i)}$ from multiple vehicles (i) for the same satellite j. The vehicle (r) must then merge these before correcting the pseudo-distance value of the satellite j measured taking into account the independence criterion of these data in order to avoid over-convergence. Rohani et al. (2016) proposed that the "DDGPS vehicle" transfers each correction information $\Delta\rho_j^{(i)}$ with the variance $\sigma_j^{(i)}$ of the GPS pseudo-distance correction, as well as the exhaustive list of the identifiers of the vehicles that took part in the calculation of this correction information. The variance value of a GPS pseudo-distance correction $\Delta\rho_j^{(i)}$ gives a confidence level to this correction $\Delta\rho_j^{(i)}$. When disjointed, the lists of identifiers linked to the correction values $\Delta\rho_j^{(i)}$ inform of the independent nature of the data. If the data are independent, then the receiving vehicle (r) merges the corrections received by means of a weighted average; it calculates the new variance (Rohani et al. 2016) and then merges the lists of identifiers. When the data are not independent, the DDGPS algorithm retains only the pseudo-distance correction $\Delta\rho_j^{(i)}$ that has the smallest value of $\sigma_j^{(i)}$.

The DDGPS application therefore requires each vehicle (i) to transmit a message containing, for each correction information, the identifiers j of the satellites; the values $\Delta\rho_j^{(i)}$, the values $\sigma_j^{(i)}$ and the list of identifiers of the vehicles that participated in the calculation of the correction $\Delta\rho_j^{(i)}$. However, the transmission of a list of identifiers with each pair of values $(\Delta\rho_j^{(i)}, \sigma_j^{(i)})$ can become bandwidth-intensive. Yet, it was stated in Chapter 2 that "each group of leaf nodes aggregated around a branch node defines a disjoint set from those formed around neighboring branch nodes." Therefore, provided that the execution of the DDGPS algorithm is restricted to branch nodes only, it is possible to avoid the transfer of vehicle identifier lists. The proposed implementation of the CMM-DDGPS collaborative localization method on the CBL-OLSR network is as follows:

– Any node, leaf and branch, can run the CMM algorithm locally and broadcast its pseudo-distance measurements within its one-hop neighborhood $\rho_j^{(i)}$ for each satellite j it sees.

– Only branch nodes run the DDGPS algorithm. Each restricts the distribution of the set of corrected pseudo-distance values and their associated variance $(\Delta\rho_j^{(i)}, \sigma_j^{(i)})$ to the group of leaf nodes that elected it and to the branch nodes of its chain located upstream of its position. By propagating them in one direction only, CBL guarantees the independence of the corrected data and the DDGPS messages then include only a set of three pieces of information for each satellite correction: j is the satellite identifier, $\Delta\rho_j^{(i)}$ is the pseudo-distance correction of this satellite and $\sigma_j^{(i)}$ is the variance associated with this correction.

4.4.4.2. *Simulation configuration*

The study was conducted using OPNET *Riverbed Modeler*. The ad hoc vehicle network consists of vehicles equipped with IEEE 802.11p cards, the configuration of which has been explained in Tables 3.16 and 4.2. The routing protocol used is the CBL-OLSR protocol from Rivoirard et al. (2018a) and Wahl et al. (2021). Two applications were modeled, the collaborative localization application (CMM-DDGPS) introduced in section 4.4.4.1 and the collective perception application previously presented in section 4.4.3. Other configuration items are as follows:

– The road network on which the VANET is located is a two-lane section of the 10 km A27 highway located between Lille and the Belgian border. The road traffic scenario chosen is the high-density S12 scenario (Table 3.12). This vehicle mobility model is derived from real traffic data collected by the French Interdepartmental Directorate of Northern Roads. The average number of vehicles per hour on the road section is shown in Figure 4.14(a).

– Vehicle positioning information is assumed to be provided only by the GPS on board the vehicles. Every vehicle is equipped with a GPS and can estimate its position, subject to an uncertainty on the calculated position, from the pseudo-distance information received from the satellites visible by its GPS receiver. The vehicles also have on-board mapping.

CMM and DDGPS messages are transmitted at fixed periods, ranging from 50 to 500 ms (with a 50-ms sampling step) depending on simulations. The number of visible satellites has been set to 12 in accordance with the study by Martineau et al. (2009). The size of the CMM and DDGPS packets was determined based on the header of a typical OLSR HELLO message. CMM packets transmitted by leaf or branch nodes are 896 bits, DDGPS packets that only branch nodes transmit are 1,280 bits.

– The collective perception application runs concurrently with the localization application in the vehicles and its transmissions are carried out following to the same transmission periodicity, according to the specificities explained in section 4.4.3.

The results presented below are obtained from several simulations ran for each of the application traffic transmission periodicities. Ten different seed values were used for the random number generator, such that the sequences of numbers drawn do not favor a particular situation during the OPNET simulations. Those results collected in terms of packet delivery ratio (equation [4.1]) and IP delay were reported as a function of the periodicity of the EPM, CMM and DDGPS messages in Figures 4.18–4.20.

4.4.4.3. *Analysis of simulation results*

Simulation results show that the maximum packet delivery ratio is reached when the interval between two transmissions is 350 ms for each network node, both for the cooperative localization application (Figure 4.17) and the collective perception application (Figure 4.18). For this transmission periodicity, the end-to-end delay achieved is around 50 ms for both applications, that is, one of the lowest (Figures 4.19 and 4.20). This 350 ms periodicity means that the transmitted information is only refreshed three times per second, which may not be sufficient for all road situations. This could be sufficient, for example, in areas where there is an enforced speed limit such as in the city, or in the presence of traffic jams, regardless of the road configuration. However, for configurations where data should be refreshed over a shorter periodicity of time, the results show that the best trade-off between short transmission periodicity and low end-to-end delay is obtained for a periodicity of 125 ms for the collaborative localization application (Figure 4.19) and 175 ms for the collective perception application (Figure 4.20). Moreover, with these transmission periods, the packet delivery ratio is close to the highest value observed with the time interval of 350 ms. The gain achieved on the end-to-end delay metric is advantageous compared to the small loss observed on the PDR. The difference observed in the values of the transmission interval of the two applications is due to the amount of traffic generated by each application. This means that road traffic conditions influence the variation of these values. As a matter of fact, when the road network is sparse (very light density) and comprises few static elements, the collective perception application should generate less traffic. It could then adopt a higher transmission interval without performance degradation. Similar reasoning can be carried out for the collaborative localization application.

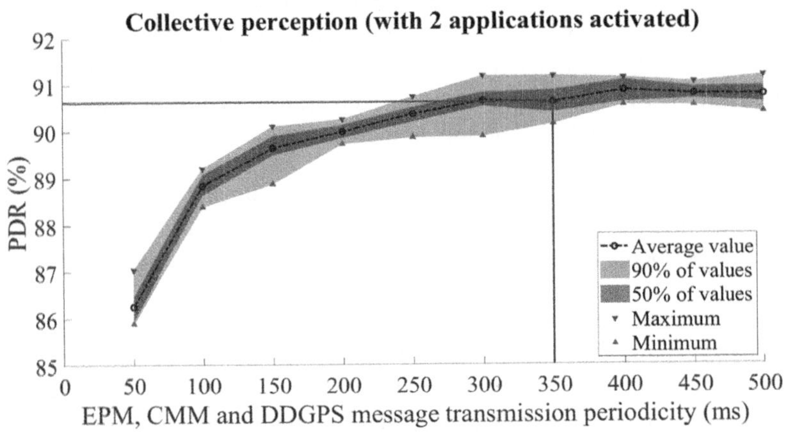

Figure 4.17. *Packet delivery ratio (PDR) obtained for the collaborative localization application when both applications are enabled. For a color version of this figure, see www.iste.co.uk/wahl/design.zip*

This study with two applications highlights that the additional delay introduced by running multiple applications simultaneously at the mobile node level is non-negligible. In comparison, the IP end-to-end delay obtained when each of the applications is performed alone is about 50 ms for the collaborative localization application with a 100-ms transmission periodicity (Rivoirard et al. 2018b). In addition, although the end-to-end delay varies only slightly for the collective perception application (Sondi et al. 2018), the PDR decreases by about 10% when concurrently running another collaborative application. These results consequently show that additional mechanisms need to be introduced in order to maintain decent performance for each individual application when they are run at the same time. Introducing prioritization mechanisms between the traffics of different collaborative applications is a solution to help improve the performance of the overall system. Another would consist of defining the evolution of the transmission rate of each application according to road traffic density and the configuration of the road network.

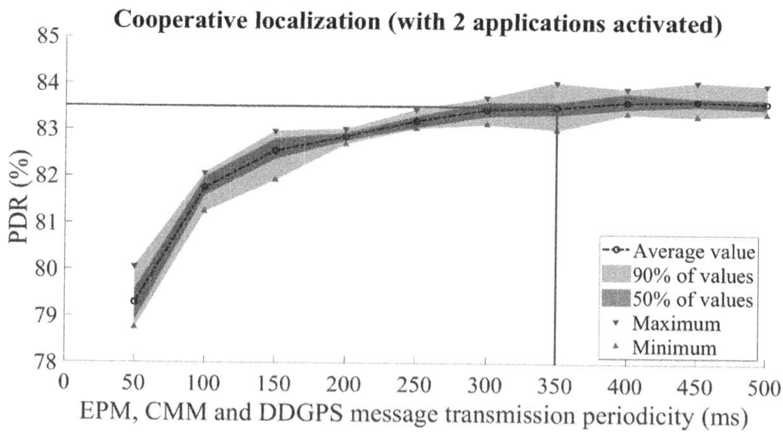

Figure 4.18. *Packet delivery ratio obtained for the collective perception application when both applications are enabled. For a color version of this figure, see www.iste.co.uk/wahl/design.zip*

4.5. Conclusion

In this chapter, structural evaluations were performed to validate the implementation of CBL in a copy of the OLSR *process model* of the OPNET network simulator *Riverbed Modeler*. The results show that the CBL-OLSR structure remains robust to changing propagation models. However, a less favorable propagation model, with the assumption of total signal obstruction beyond 210 m in the presence of an obstructing vehicle, reduces the stability of the leaf–branch and branch–branch links, which could disturb the broadcasting of TC messages in a sparse network or conversely in a high-density one.

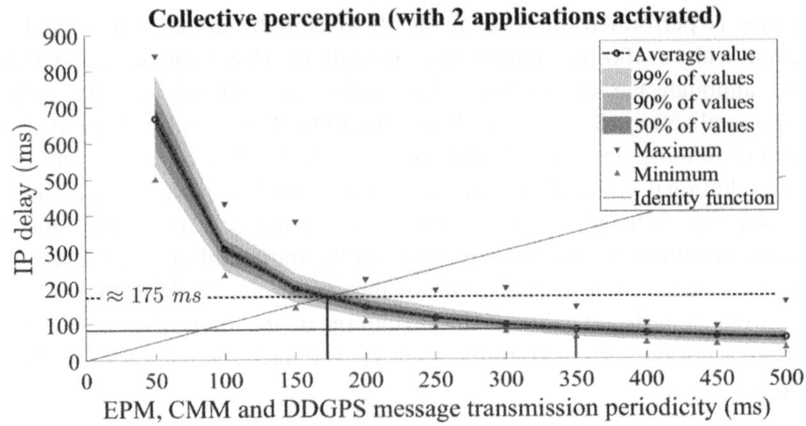

Figure 4.19. *IP end-to-end delay in milliseconds obtained for the collaborative localization application when both localization and perception applications are enabled. For a color version of this figure, see www.iste.co.uk/wahl/design.zip*

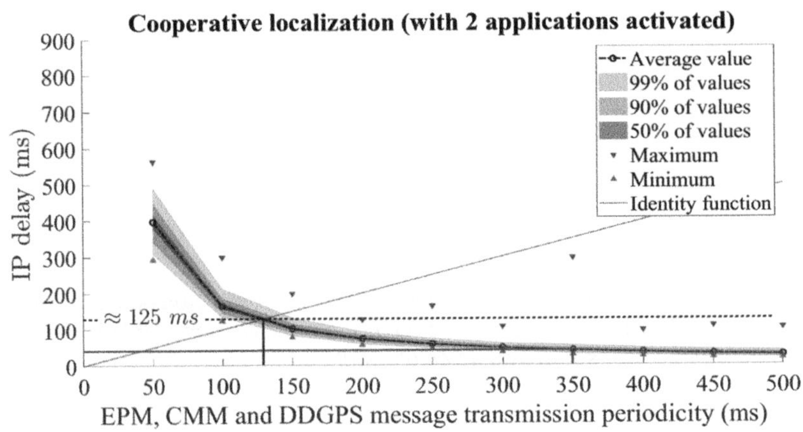

Figure 4.20. *IP end-to-end delay in milliseconds obtained for the collective perception application when both localization and perception applications are enabled. For a color version of this figure, see www.iste.co.uk/wahl/design.zip*

A comparative study of routing traffic generated by CBL-OLSR and native OLSR with five different highway scenarios shows a decrease in routing load with CBL-OLSR due to less TC message traffic generated. This decrease, which is all the greater for higher road densities, is due to a lower number of branch nodes in the CBL-OLSR network compared to MPR nodes in the OLSR network. Collaborative applications make it possible to exploit the potential of ad hoc networks for driver assistance and automatic vehicle steering. A large number of propositions have been submitted over the last 5 years in order to design accurate and reliable applications for collaborative localization or collective perception based on cooperation between vehicles through direct, V2V exchanges of data necessary for the operation of these collaborative applications. The evaluation of the two applications simultaneously deployed in various scenarios shows that for relatively high vehicle density, for which each vehicle has about 4–15 direct neighbors, each application can operate with a 150-ms transmission periodicity while keeping a packet delivery ratio above 80% and an IP end-to-end delay of about 150 ms.

4.6. References

Abbas, T., Sjöberg, K., Karedal, J., Tufvesson, F. (2015). A measurement based shadow fading model for vehicle-to-vehicle network simulations. *International Journal of Antennas and Propagation*, 190607, 12. doi: 10.1155/2015/190607.

Al-Ezaly, E., Abo-Elfetoh, A., Elhishi, S. (2021). An optimized location service for the fifth generation vanets inspired by traffic lights. *PLOS ONE*, 16(11), 1–17. doi: 10.1371/journal.pone.0259060.

Ashok, D.M., Pai, M.M.M., Mouzna, J. (2011). Efficient map based location service for VANETs. In *2011 11th International Conference on ITS Telecommunications*. St. Petersburg, 387–392. doi: 10.1109/ITST.2011.6060087.

Chaabouni, N., Hafid, A.S., Rezgui, J., Cherkaoui, S. (2016). Setting up an extended perception in a vehicular network environment: A proof of concept. In *2016 IEEE Wireless Communications and Networking Conference*, Doha, 1–7. doi: 10.1109/WCNC.2016.7564766.

Chen, R., Jin, W.-L., Regan, A. (2010). Broadcasting safety information in vehicular networks: Issues and approaches. *IEEE Network*, 24(1), 20–25. doi: 10.1109/mnet.2010.5395779.

Clausen, T., Jacquet, P., Adjih, C., Laouiti, A., Minet, P., Muhlethaler, P., Qayyum, A., Viennot, L. (2003). Optimized link state routing protocol (OLSR). Document, RFC 3626. doi: 10.17487/RFC3626.

Dar, K., Bakhouya, M., Gaber, J., Wack, M., Lorenz, P. (2010). Wireless communication technologies for ITS applications [topics in automotive networking]. *IEEE Communications Magazine*, 48(5), 156–162. doi: 10.1109/mcom.2010.5458377.

ETSI (2009). Intelligent transport systems (ITS); Vehicular communications; Basic set of applications; Definitions. Technical Report ETSI TR 102 638 V1.1.1 (2009-06), ETSI Technical Committee Intelligent Transport System, Sophia Antipolis [Online]. Available at: https://www.etsi.org/standards.

ETSI (2012). Intelligent transport systems (ITS); STDMA recommended parameters and settings for cooperative ITS; Access layer part. Technical Report ETSI TR 102 861 V1.1.1 (2012-01), ETSI Technical Committee Intelligent Transport System, Sophia Antipolis [Online]. Available at: http://www.etsi.org/deliver/etsi_tr/102800_102899/102861/01.01.01_ 60/tr_102861v010101p.pdf.

ETSI (2013a). Intelligent transport systems (ITS); V2C applications; Part 1: Road hazard signalling (RHS) application requirements specification. Technical Specification ETSI TS 101 539-1 V1.1.1 (2013-08), ETSI Technical Committee Intelligent Transport System, Sophia Antipolis [Online]. Available at: https://www.etsi.org/standards.

ETSI (2013b). Intelligent transport systems (ITS); V2X applications; Part 3: Longitudinal collision risk warning (LCRW) application requirements specification. Technical Specification ETSI TS 101 539-3 V1.1.1 (2013-11), ETSI Technical Committee Intelligent Transport System, Sophia Antipolis [Online]. Available at: https://www.etsi.org/standards.

ETSI (2014a). Intelligent transport systems (ITS); Vehicular communications; Basic set of applications; Local dynamic map (LDM). European Standard ETSI EN 302 895 V1.1.1 (2014-09), ETSI Technical Committee Intelligent Transport System, Sophia Antipolis [Online]. Available at: http://www.etsi.org/standards, pages = 36.

ETSI (2014b). Intelligent transport systems (ITS); Vehicular communications; Basic set of applications; Part 2: Specification of cooperative awareness basic service. European Standard ETSI EN 302 637-2 V1.3.2 (2014-11), ETSI Technical Committee Intelligent Transport System, Sophia Antipolis [Online]. Available at: https://www.etsi.org/standards.

ETSI (2018). Intelligent transport systems (ITS); V2X applications; Part 2: Intersection collision risk warning (ICRW) application requirements specification. Technical Specification ETSI TS 101 539-2 V1.1.1 (2018-06), ETSI Technical Committee Intelligent Transport System, Sophia Antipolis [Online]. Available at: https://www.etsi.org/standards.

ETSI (2019a). Intelligent transport systems (ITS); Vehicular communications; Basic set of applications; Analysis of the collective perception service (CPS); Release 2. Technical Report ETSI TR 103 562 V2.1.1 (2019-12), ETSI Technical Committee Intelligent Transport System, Sophia Antipolis [Online]. Available at: https://www.etsi.org/standards.

ETSI (2019b). Intelligent transport systems (ITS); Vehicular communications; Basic set of applications; Part 2: Specification of cooperative awareness basic service. European Standard ETSI EN 302 367-2 V1.4.1 (2019-04), ETSI Technical Committee Intelligent Transport System, Sophia Antipolis [Online]. Available at: https://www.etsi.org/standards.

ETSI (2019c). Intelligent transport systems (ITS); Vehicular communications; Basic set of applications; Part 3: Specifications of decentralized environmental notification basic service. European Standard ETSI EN 302 367-3 V1.3.1 (2019-04), ETSI Technical Committee Intelligent Transport System, Sophia Antipolis [Online]. Available at: https://www.etsi.org/standards.

Fallgren, M. and Timus, B. (2013). Scenarios, requirements and KPIs for 5G mobile and wireless system: Mobile and wireless communications Enablers for the Twenty-twenty Information Society (METIS). Document, ICT-317669-METIS/D1.1, 7th Framework Programme (FP7) [Online]. Available at: https://cordis.europa.eu/docs/projects/cnect/9/317669/080/deliverables/001-METISD11v1pdf.pdf.

Giang, A.T. (2014). Capacity of vehicular Ad-hoc NETwork. PhD Thesis, Université Paris Sud, Paris [Online]. Available at: https://tel.archives-ouvertes.fr/tel-00989836/.

Günther, H.-J. (2018). Collective perception in vehicular ad-hoc networks. PhD Thesis, Technische Universität Braunschweig, Braunschweig. doi: 10.24355/dbbs.084-201802021049.

Günther, H.-J., Trauer, O., Wolf, L. (2015). The potential of collective perception in vehicular ad-hoc networks. In *2015 14th International Conference on ITS Telecommunications (ITST)*, 1–5. IEEE, Copenhagen. doi: 10.1109/itst.2015.7377190.

Idrac, A.-M. (2018). Développement des véhicules autonomes – orientations stratégiques pour l'action publique. Report, Ministère de la transition écologique et solidaire, Paris [Online]. Available at: http://www.ladocumentationfrancaise.fr/Reports-publics/184000287/index.shtml.

Kaplan, E. and Hegarty, C. (2005). *Understanding GPS: Principles and Applications*. Artech House, Norwood.

Laouti, A., Mühlethaler, P., Najid, A., Plakoo, E. (2002). Simulation results of the OLSR routing protocol for wireless network. Research Report RR-4414, INRIA, Paris [Online]. Available at: https://hal.inria.fr/inria-00072174/file/RR-4414.pdf.

Martineau, A., Macabiau, C., Mabilleau, M. (2009). GNSS RAIM assumptions for vertically guided approaches. In *GNSS 2009, 22nd International Technical Meeting of The Satellite Division of the Institute of Navigation*, Savannah, 2791–2803 [Online]. Available at: https://hal-enac.archives-ouvertes.fr/hal-01022160/.

Naimi, A. and Jacquet, P. (2003). Le contrôle du délai dans le protocole OLSR. Research Report RR-4894, INRIA, Paris [Online]. Available at: https://hal.inria.fr/inria-00071688.

Paret, D. and Rebaine, H. (2019). *Véhicules autonomes et connectés. Techniques, technologies, architectures et réseaux : du multiplex à l'ethernet automobile*. Dunod, Paris.

Patel, K.S. and Shah, J.S. (2015). Analysis of AODV for finding out ideal queue size. In *2015 International Conference on Computer, Communication and Control (IC4)*, Indore, 1–3. doi: 10.1109/IC4.2015.7375730.

Perkins, C., Belding-Royer, E., Das, S. (2003). Ad-hoc On-Demand Distance Vector (AODV) Routing. Document, RFC 3561. doi: 10.17487/RFC3561.

Pollard, E. and Gingras, D. (2012). Improved low cost GPS localization by using communicative vehicles. In *2012 12th International Conference on Control Automation Robotics & Vision (ICARCV)*, 317–322. IEEE, Guangzhou. doi: 10.1109/icarcv.2012.6485178.

Rivoirard, L. (2018). Modèle d'auto-organisation pour les protocoles de routage dans les réseaux ad hoc de véhicules : application à la perception élargie et à la localisation coopératives. PhD Thesis, Université de Lille, Lille [Online]. Available at: http://www.theses.fr/2018LIL1I041.

Rivoirard, L., Wahl, M., Sondi, P., Berbineau, M., Gruyer, D. (2018a). Chain-Branch-Leaf: A clustering scheme for vehicular networks using only V2V communications. *Ad Hoc Networks*, 68, 70–84. doi: 10.1016/j.adhoc.2017.10.007.

Rivoirard, L., Wahl, M., Sondi, P., Gruyer, D., Berbineau, M. (2018b). A cooperative vehicle ego-localization application using V2V communications with CBL clustering. In *2018 IEEE Intelligent Vehicles Symposium 4*, Changshu, 722–727. doi: 10.1109/ IVS.2018.8500688.

Rohani, M., Gingras, D., Gruyer, D. (2016). A novel approach for improved vehicular positioning using cooperative map matching and dynamic base station DGPS concept. *IEEE Transactions on Intelligent Transportation Systems*, 17(1), 230–239. doi: 10.1109/tits.2015.2465141.

Rondinome, M. and Correa, A. (2019). D5.1. Definition of V2X message sets. Delivery of Project Number "Horizon 2020 ART-05-2016 – GA Nr 723390 ART-TransAID" (Transition Areas for Infrastructure-Assisted Driving). WP5 Connectivity and Signalling [Online]. Available at: https://www.transaid.eu/deliverables/.

Sheu, J.-P., Hu, W.-K., Bol, R.E. (2012). A registration system for aiding in localization and routing in hybrid VANETs. In *2012 12th International Conference on ITS Telecommunications*, Taipei, 694–699. doi: 10.1109/ITST.2012.6425271.

Shimada, H., Yamaguchi, A., Takada, H., Sato, K. (2015). Implementation and evaluation of local dynamic map in safety driving systems. *Journal of Transportation Technologies*, 5(02), 102–112. doi: 10.4236/jtts.2015.52010.

Shuttleworth, J. (2019). SAE standards news: J3016 automated-driving graphic update [Online]. Available at: https://www.sae.org/news/2019/01/sae-updates-j3016-automated-driving-graphie [Accessed 19 January 2021].

Sondi, P., Gantsou, D., Lecomte, S. (2014). A multiple-metric QoS-aware implementation of the optimised link state routing protocol. *International Journal of Communication Networks and Distributed Systems*, 12(4), 381–400. doi: 10.1504/ IJCNDS.2014.062227.

Sondi, P., Rivoirard, L., Wahl, M. (2018). Performance evaluation of vehicle-to-vehicle communications for a collective perception application in vehicular ad hoc networks. In *2018 IEEE 29th Annual International Symposium on Personal, Indoor and Mobile Radio Communications (PIMRC)*, Bologna, 602–603. doi: 10.1109/PIMRC.2018.8580753.

Wahl, M., Sondi, P., Rivoirard, L. (2021). Enhanced CBL clustering performance versus GRP, OLSR and AODV in vehicular Ad Hoc networks. *Telecommunication Systems*, 76, 525–540. doi: 10.1007/s11235-020-00734-1.

Zhiyuan, L. (2009). Geographic routing protocol and simulation. In *Second International Workshop on Computer Science and Engineering (WCSE), volume 2*, 404–407. IEEE, Qingdao. doi: 10.1109/WCSE.2009.840.

5

Formal Model for the Analysis of the Properties of the Protocol

5.1. Introduction

Wireless communications technologies make one of the main supports for the development of the future Internet (connected objects, vehicles and cities). The convergence of network architectures toward all-IP has led to the emergence of the Internet of Things and makes possible the opportunistic operation of connected objects that can communicate by way of an infrastructure when it is operational and switch to ad hoc mode when it is not. Ad hoc networks, in particular, ensure great flexibility by enabling the possibility of short-range communications between all connected nodes, even in a context deprived of any preexisting communication infrastructure. The inherently decentralized coordination of ad hoc networks makes their security, robustness and performance highly dependent on the protocols used to ensure the discovery of the objects evolving in the network, their self-organization, routing, quality of service (QoS) and security. The design and evaluation of communication protocols is usually based on a functional model built according to the needs of the system. We illustrated this approach in Chapter 2 on an ad hoc routing protocol recently proposed for the road context. Simulation-based studies on these protocols, such as those described in Chapters 3 and 4, often focus on measuring the performance obtained under specific deployment scenarios.

Yet, the general trend toward the development of autonomous vehicles in public and individual transport today emphasizes the need for robust communication protocols with guaranteed properties, in particular in terms of QoS, operational safety and security. Thereby, several works have been undertaken for the development of methodologies for the design and evaluation of communication protocols that follow this avenue by relying on the methods and tools provided by formal modeling. This chapter presents the contributions we have developed to this end.

After a brief overview of formal methods and tools as well as a brief review of some previous work applying them to ad hoc routing protocols, this chapter describes their application to the protocol of the case study. Through the description of the models developed, this chapter shows how this design and evaluation approach complements those described in the previous three chapters. The research perspectives opened up by this research topic conclude the presentation of this chapter.

5.2. The contribution of formal methods to protocol engineering

The reliability of critical systems can be improved through the use of formal methods for developing their critical subsystems (Rushby 1993). Formal methods can be applied at different stages of the development process, including during code specification, verification and generation. Mathematical proofs serve as a means of ensuring that the specifications meet the requirements. The evolution of automatic generation of executable code directly from formal specifications reduces the need for unit testing, as the correctness of the build is guaranteed by mathematical proofs. The complete formalization of a system has often been pointed out as a difficult and expensive task (Bowen and Hinchey 2006). Various lightweight formal methods (Jackson 2001; Barner 2005) have been proposed for performing partial formal specifications of certain functionalities in applications. As examples, a collection of works can be mentioned (Cavalli 2001) listing formal languages for QoS specification, and in particular E-LOTO (Leduc and Kuty 2001), and formal tools such as Estelle. Similar studies were proposed for the verification of security properties in ad hoc routing protocols (Lee 2014; Diaz et al. 2014). An up-to-date review of the research developed in this topic can be found, in particular, in Fakhfakh et al. (2019) and Chebbi (2019).

The most cited tools in this work include the B-Method (Abrial 2005), which is often used for reasoning complex systems. An event-driven extension of the B-Method, Event-B, relies on the notion of event for describing actions rather than operations, which makes it very useful for modeling and analyzing routing protocols that are essentially event-based software programs. A formal analysis of the functioning of the DSR (*Dynamic Source Routing*) protocol with Event-B was provided by Méry and Singh (2011). Similar work was proposed, respectively, on the AODV (*Ad hoc On-demand Distance Vector*) protocol (Singh and Singh 2013), on the OLSR (*Optimized Link State Routing*) protocol (Kamali and Petre 2016) and more recently on the ZRP (*Zone Routing Protocol*) (Fu and Zheng 2019). These analyses focused on the basic functionalities of these protocols, namely, the detection of links between nodes in the network, route discovery and relay node selection.

Nevertheless, a major feature of ad hoc routing protocols in the context of vehicular networks is the realization of network self-organization. It requires the spatial relationships between vehicles as well as their movements in the road environment to be taken into consideration. In all of the existing approaches reported above, the integration of this spatial dimension and mobility is not achieved in the formal model. An important contribution of our (Chebbi et al. 2019) work on the application of formal methods to the design and evaluation of ad hoc routing protocols consisted of integrating this aspect into the analysis of the case study protocol. This chapter also describes how the concepts and functionalities of the case study ad hoc routing protocol can be formally described and analyzed with Event-B.

5.3. The Event-B method

Event-B (Abrial 2010) is a formal modeling method based on solid mathematical foundations, in this case set theory and first-order logic. A detailed description of the method would fall beyond the scope of this chapter. Nevertheless, sufficient insight is required to acquire an understanding of the models presented therein.

5.3.1. *Mathematical concepts and notations of the Event-B method*

Event-B mathematical notations necessary for the description of the models that will be presented in this chapter can be listed as follows:

– $A \times B$ denotes the set of tuples of the form $x \mapsto y$, where $x \in A$ and $y \in B$;

– $A \subseteq B$ means that A is a subset of B (all elements of A are included in B);

– $R \in A \leftrightarrow B$ expresses that R is a definite relation from A to B ($R \subseteq A \times B$);

– $dom(R)$ (respectively, $ran(R)$) is a subset of A (respectively, B) having its images (respectively its antecedents) in B (respectively in A) by the relation R;

– $f \in A \pto B$ expresses that f is a definite partial function from A to B;

– $f \in A \pinj B$ expresses that f is a partial injective function defined from A to B. f is a function that preserves distinction;

– $f \in A \rightarrow B$ expresses that f is a total function defined from A to B and whose domain is A $dom(f) = A$;

– $card(A)$ denotes the number of elements in the finite set A;

– $A \lhd id$ denotes the set of tuples (a, a) where $a \in A$;

– $A \setminus B$ denotes the set of elements of A that do not belong to B;

– $X \lhd R$ denotes the restriction of the domain of a relation R to X;

– $X \ndres R$ denotes the anti-restriction of the domain of a relation R to X;

– $R \rhd Y$ denotes the restriction of the codomain of a relation R to Y;

– $R \rhd Y$ denotes the anti-restriction of the codomain of a relation R to Y;

– $Q \Leftarrow R$ denotes the overloading of Q by R defined as:

$$Q \Leftarrow R = R \cup \{x \mapsto y | x \mapsto y \in Q \wedge x \notin dom(R)\}$$

Other common mathematical notations (e.g. $\in$, $\notin$ and $\mathbb{N}$) will be used, although their definition is not formally proposed in this chapter.

5.3.2. *Event-B model structure*

As described in Figure 5.1, an Event-B model relies on two types of entities for the description of a system: *machine* and *context*. A *context* represents the static part of the model and contains the following clauses:

– `SETS`: used to define abstract types and enumerated types;

– `CONSTANTS`: used to define constant elements;

– `AXIOMS`: used to express static properties on constants and sets.

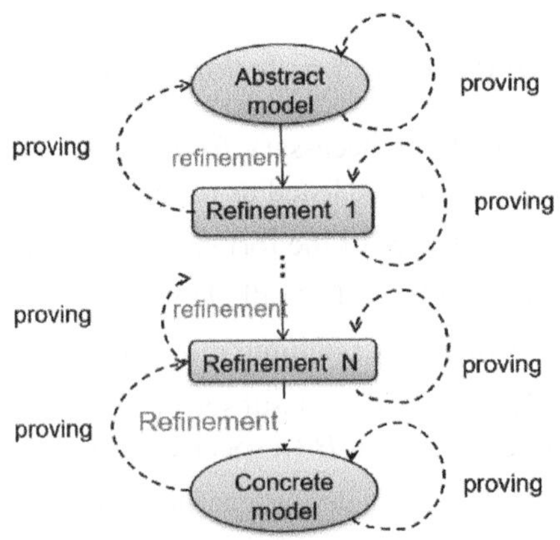

Figure 5.1. *Correction-by-construction methodology and Event-B model structure. For a color version of this figure, see www.iste.co.uk/wahl/design.zip*

A *machine* represents the dynamic part of a model. It can "see" one or more *context*, and is also structured into several clauses:

– `SEES`: used to include the list of *context* that will be used by the machine;

– `VARIABLES`: used to define a set of variables;

– `INVARIANTS`: used to define variable types, as well as model properties and rules;

– `EVENTS`: used to define the set of events of the model. This clause must *at least* include the machine initialization event.

An event describes a change in the state of variables on the machine in which they are defined. It can take as input a set of local parameters. These parameters, along with the conditions for triggering an event, are called *guards*. The behavior of an event is described by a set of substitution actions whose general form is described as follows:

$$a : |P(a, a')$$

This formulation expresses the relationship between the value of the variable a before and after the substitution (a').

5.3.3. *Refinement concept*

The effectiveness of the Event-B method is essentially based on the concept of refinement. The latter makes it possible to decompose the formulation of a model for a complex system with different abstraction levels. As shown in Figure 5.1, designers start with an abstract requirement specification. They then refine this specification in successive steps to obtain a detailed description of the system, which can be mapped to executable code (in C, JAVA, ADA, etc.) by means of tools such as EB2C (Méry and Singh 2010). It is quite possible to apply refinement to both *context* and *machine*. A *context* can be extended by adding new sets, constants and/or axioms. A *machine* can be refined by adding or replacing variables with new variables. New events can also be added to implicitly refine an event called *skip event*.

5.3.4. *Event-B model correction*

Correctness of an Event-B model is ensured through *proof obligations* (POs). POs are automatically generated by the formal tool; in this case, Rodin was used for the case study. Different types of POs are generally defined: preservation invariants (INVs) and abstract guards strengthenings (GRDs).

`INV`-type POs ensure that the execution of any event must preserve the related invariant in the machine where it is defined. Therefore, for each event of the form $(G \mid Act)$, it must be proved that:

$$(G \wedge I \Rightarrow [Act]I)$$

where $[Act]I$ provides the weakest precondition on the *before* state such that executing action Act leads to an *after* state that satisfies invariant I. This PO ensures that every event performed in a given state satisfying the invariant leads to a state in which the same invariant is also satisfied.

GRD-type POs ensure that the *guard* of a refined event is more restrictive than that of the corresponding abstract event. This ensures that when a concrete event is activated, the corresponding abstract event is also activated. Let A (respectively, B) be a machine comprising an invariant I_A (respectively, I_B) and an event E (respectively, R) comprising a *guard* G_E (respectively, G_R). Event R correctly refines event E if and only if the following evidence is obtained:

$$I_A \wedge I_B \wedge G_R \Rightarrow G_E$$

In order to relieve the various POs, the Rodin platform offers a series of automatic provers such as SMT and Atelier B used in the context of our work. The Rodin tool also includes mechanisms for interacting with provers and providing expert knowledge likely to trigger well-directed inferences.

5.4. Formal approach applied to the protocol of the case study

As in most grouping methods, *Chain-Branch-Leaf* (CBL) (Rivoirard et al. 2018) sets out several rules that must be applied for each vehicle in order to allow the construction of groups and the backbone according to the desired organization model. For example, the uniqueness of the branch node in each group formed, attaching each leaf node to a single branch node at each time instant, etc., are all rules that guarantee the properties of the structure formed by CBL. Nonetheless, without a formal description of the protocol, the effects of these rules on these properties can be merely approached, but not formally verified. As such, the analysis of the CBL protocol involves two steps. The first step consists of the validation of the rules, which verifies that the protocol specification complies with the functional requirements defined by the designer (Zowghi and Gervasi 2004), in this case in Chapter 2 for the case study protocol. The second step consists of verifying the consistency of the protocol rules to ensure that they do not introduce contradictions.

The implementation of these steps with Event-B will result in obtaining, through a series of refinements, a correct-by-construction model of the CBL protocol (Sondi et al. 2021). In this formal methods-based design approach, evaluation is an integral part of the design process. The formal tool will actually make it possible to verify and prove the correctness of the CBL model at the same time as its operating rules and its considered properties. Therefore, when the formal model of the CBL protocol is verified, it guarantees both its consistency and operation-free functional errors.

5.4.1. *Formal description of the organization by CBL in a VANET*

Without going over a detailed description of the functioning of CBL already proposed in Chapter 2, this section reviews the key concepts and steps through the presentation of their transcription in the formal model proposed for the protocol. The basic assumptions for the operation of CBL are therefore assumed to be established, namely:

– CBL is a distributed algorithm run independently by each vehicle, which is supposed to have an interface that gives it access to full knowledge of its position at all times (GPS, for example).

– Vehicle mobility is carried out according to the actual constraints imposed by the configuration of the roads. Several configurations are described in Chapters 3 and 4. Schematically, on a two-lane freeway, CBL creates two backbones, one for each direction of traffic (Figure 5.2).

– Two types of nodes are defined, namely branch nodes that are cluster-heads and the only allowed to retransmit traffic across the network; and leaf nodes that elect a single branch node to which each forwards its traffic destined for the other nodes of the network.

– The chain or backbone is formed by the selected branch nodes in the same traffic direction of the road infrastructure.

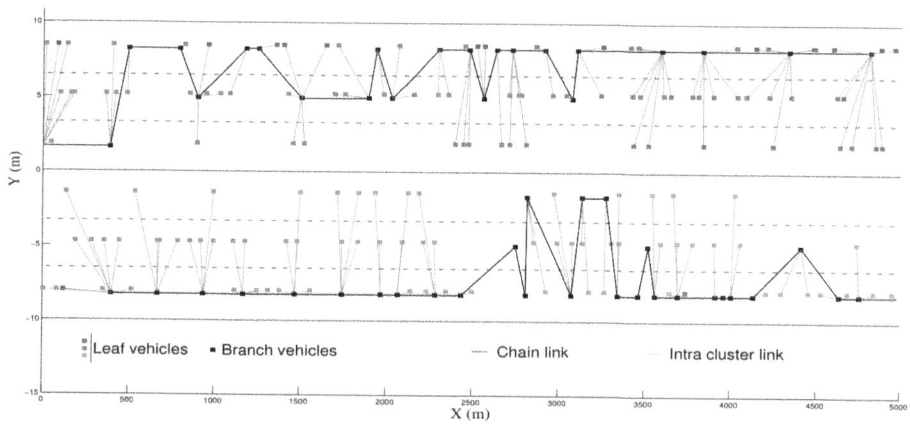

Figure 5.2. *Example of CBL chains on a two-lane highway. For a color version of this figure, see www.iste.co.uk/wahl/design.zip*

The set of nodes in the vehicle network is denoted $Nodes$. Without loss of generality, this description of the protocol description will only take two types of packets into account. On the one hand, there are the so-called control packets (or *Hello packets*) employed for signaling and discovering network nodes. On the other hand, packets containing application flows. The set of all control packets transmitted by the nodes in the network is denoted $Hello$. The links between the vehicles in the network are formally defined by the following functions:

$$links \in \mathbb{P}(Nodes \times Nodes) \quad [5.1]$$

$$links \cap (Nodes \lhd id) = \varnothing \quad [5.2]$$

The existence of an element $n1 \mapsto n2$ in the $links$ set ($n1 \mapsto n2 \in links$) reflects the fact that the nodes $n1$ and $n2$ have established a symmetric link, according to the discovery process presented in Chapter 2. The local (or one-hop) neighborhood of a node is formally defined by the following function:

$$neighbors \in Nodes \rightarrow \mathbb{P}(Nodes) \quad [5.3]$$

For any node n, the following property must be verified:

$$\{n\} \times neighbors(n) \subseteq \{n\} \lhd links \quad [5.4]$$

The position of the vehicle represents a crucial information for the operation of the CBL protocol. As mentioned in the assumptions, each vehicle is assumed to be equipped with a global positioning device such as GPS or Galileo. However, in order to comply with the purely ad hoc nature of the network, any use of a terrestrial infrastructure for the purposes of localization can only be considered on an optional basis, and in no way conditions the operation of CBL. The position information of a node is periodically transmitted in the form of a two-dimensional vector to all of its direct neighbors via HELLO packets. The position of nodes is formally described by the function:

$$positionTable \in Nodes \rightarrow (Nodes \nrightarrow (\mathbb{N} \times \mathbb{N})) \quad [5.5]$$

Considering a node n, the set $positionTable(n)$ contains the position of said node n as well as that of its direct neighbors. It is thus possible to define the relative position of a node relatively to each of its neighbors moving in the same traffic flow direction. As it happens, each node can be located downstream or upstream of each of its direct neighbors. These relative positions can be formally defined by the following functions:

$$down \in Nodes \rightarrow \mathbb{P}(Nodes) \quad [5.6]$$

$$up \in Nodes \rightarrow \mathbb{P}(Nodes) \quad [5.7]$$

The following operator can be used to deduce the relative position of any two nodes relatively to one another:

$$DIFF = (\lambda(a \mapsto b) \mapsto (c \mapsto d) \cdot \{a, b, c, d\} \subseteq \mathbb{N} | (a - c))$$

Consider $n1$ and $n2$ two neighboring nodes, as well as their respective positions p1 and p2. The relative positions can be determined as follows:

– $n2$ is an upstream neighbor of $n1$ if $DIFF(p1 \mapsto p2)$ is positive.

– $n2$ is a downstream neighbor of $n1$ if $DIFF(p1 \mapsto p2)$ is negative.

When a node no longer receives HELLO messages from one of its neighbors for a certain period of time, the neighbor is considered to have lost its connection to the original node that can legitimately remove it from its neighbor table. This period of time is identified as *neighbor expiry time* and is formally defined as follows:

$$Neighbor\_expiry\_time \in \mathbb{N} \qquad [5.8]$$

At any given time, each node can only be a leaf or a branch. This status is formally described as follows:

$$hasType \in Nodes \rightarrow (Nodes \nrightarrow \{0, 1\}) \qquad [5.9]$$

A branch node n ($hasType(n)(n) = 1$) is a cluster-head elected by other nodes in its local neighborhood. A leaf node n ($hasType(n)(n) = 0$) is an ordinary node that associates with the branch node of the cluster to which it belongs. The chain is formed by the branch nodes moving through the road infrastructure in the same direction of travel. It is formally defined as follows:

$$chainUP \in Nodes \rightarrow \mathbb{P}(Nodes \times Nodes) \qquad [5.10]$$

$$chainDO \in Nodes \rightarrow \mathbb{P}(Nodes \times Nodes) \qquad [5.11]$$

These two functions are semantically opposed. Actually, $chainUP$ (respectively $chainDO$) refers to the upstream (respectively downstream) chain of any branch node in the chain. If branch node b2 is an upstream branch node of a branch node b1, then the latter is the downstream branch node of branch node b2. The association of each leaf node with a branch node is formally described by the following function:

$$branchChoice \in Nodes \rightarrow \mathbb{P}(Nodes \times Nodes) \qquad [5.12]$$

For any node n, the variables branchChoice(n) and chainUP(n) (respectively, chainDO(n)) refer, respectively, to the elected branch node, when the electing node is a leaf node, and to the upstream (respectively downstream) branch node, when the electing node is a branch node. They are included in the neighbor table of node n.

When a node issues a HELLO packet, it includes several data statistics about its local variables. In addition to its own identity as a sender, its position and type (branch node or leaf node) are included and, depending on the latter, its choice of branch node or its choice of upstream and downstream branch nodes is also included. The functions associated with these data are formally defined as follows:

$$helloPosition \in Hello \nrightarrow (\mathbb{N} \times \mathbb{N}) \quad [5.13]$$

$$helloBrChoice \in Hello \nrightarrow Nodes \quad [5.14]$$

$$helloChainUP \in Hello \nrightarrow Nodes \quad [5.15]$$

$$helloChainDO \in Hello \nrightarrow Nodes \quad [5.16]$$

$$helloSrType \in Hello \nrightarrow \{0, 1\} \quad [5.17]$$

Therefore, when a node n sends a HELLO packet referred to as h, the data included therein, namely, $helloPosition(h)$, $helloBrChoice(h)$, $helloChainUP(h)$, $helloChainDO(h)$ and $helloSrType(h)$, represent the list of positions, the choice of branch node, the choices of the upstream and downstream branch nodes, the type of node n itself and those of its neighbors, respectively. All of these data are essential for the CBL protocol to properly function.

A final determining parameter in the organization of CBL is the connection time. It corresponds to the estimation made at any time of the duration that an operational link is maintained between two nodes according to their respective movements. Used in particular when designating branch nodes by leaf nodes, it is formally defined as follows:

$$Ctime \in (Nodes \times Nodes) \nrightarrow \mathbb{N} \quad [5.18]$$

5.4.2. *CBL rules and properties in a VANET*

Based on this formal description of the organization put in place by CBL in the vehicle network, it is now simpler to propose a formal expression of the operating rules and properties of the protocol as follows:

– *REQ 1*: a node with no neighbors must have the status of a leaf node;

– *REQ 2*: only one leaf node can participate in the election of a branch node;

– *REQ 3*: self-election as a branch node is prohibited;

– *REQ 4*: any branch node can only have at most one downstream branch node;

– *REQ 5*: any branch node can have at most only one upstream branch node;

– *REQ 6*: every branch node must have been elected by another node at least;

– *REQ 7*: a node designating a downstream branch node must be of the branch type;

– *REQ 8*: a node designating an upstream branch node must be of the branch type;

– *REQ 9*: if node $n1$ is a branch node downstream of node $n2$, then the latter is a branch node upstream of node $n1$;

– *REQ 10*: the upstream branch node of a node must be designated from the neighbors located upstream of the node;

– *REQ 11*: the downstream branch node of a node must be designated from the neighbors downstream of the node;

– *REQ 12*: one of the neighbors of a leaf node must be elected as a branch.

The rules applied specifically during election operations of a branch node or an upstream/downstream branch node (NER) are as follows:

– *NER1*: self-election is prohibited (the electing node must be different from the elected node);

– *NER2*: an elected node must be a neighbor of its electing node (a node can only elect one of its neighbors);

– *NER3*: when a branch node elects a node to extend the chain, it must choose it from its upstream neighbors. This rule makes it possible to restrict the extension of the chain in one direction to reduce the risk of multiple chains (Wahl et al. 2021);

– *NER4*: a branch node must have no more than one upstream branch node;

– *NER5*: when a leaf node has no branch nodes among its neighbors, it must elect as its branch node the one among its neighbors capable of providing the highest connection time (Ctime);

– *NER6*: when a branch node has no upstream branch nodes, it must choose as its upstream branch node the one among its neighbors that provides the highest connection time (Ctime).

All nodes that join the vehicular network are initially leaf nodes. Some of them end up changing their status and adopting the status of branch node. Given that movements take place in the vehicular network, any node is likely to change its status again. The following rules (TCR) set out the conditions for these changes:

– *TCR1*: when a leaf node is elected by another leaf node, it must then adopt the status of branch node.

– *TCR2*: when the electing node is a branch node, then it must be selected as the downstream branch node by the elected node.

– *TCR3*: any branch node that overtakes its downstream branch node in the direction of road traffic must change its own status to a leaf node.

5.5. A correct-by-construction model of CBL with Event-B

This section initiates the description of the Event-B model itself, which has been proposed to formally transcribe the operating rules and properties of the case study protocol, namely, CBL. Figure 5.3 summarizes the architecture of the resulting model. Three levels of abstraction comprise the architecture, the first two levels of which can be reused to formally model any other communication protocol.

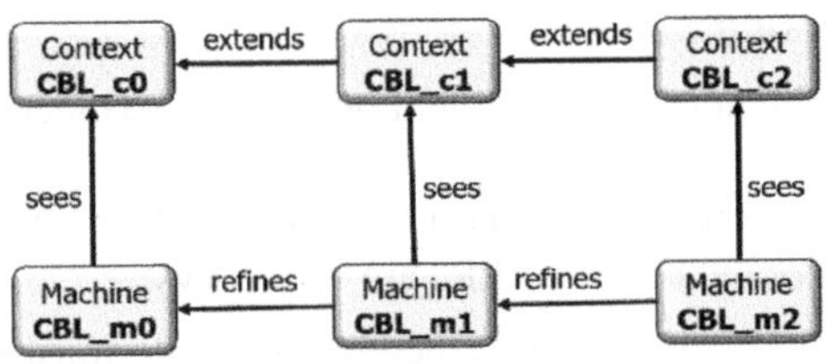

Figure 5.3. *Event-B model architecture of the CBL protocol*

5.5.1. *Level 1: basic model for any routing protocol*

This model, which provides the basic functionality needed for any routing protocol, is largely inspired by the proposition of Kamali and Petre (2016), also achieved with Event-B. It includes set definitions of network nodes, links and events related to packet operations (Figure 5.4). This model can be extended or refined in order to model any other routing protocol for wired or wireless networks. Although the substance of the model is more or less the same, adjustments have been introduced during its transcription in order to facilitate the understanding of the refinements we propose. The model is implemented by a context *CBL_c0* and a machine *CBL_m0,* as shown in Figure 5.3.

The machine CBL_m0 is said to "see" CBL_c0 (word *SEES*, Figure 5.5). The context CBL_c0 describes the static aspects of a routing protocol through the following elements:

– $Nodes$: a finite set of vehicles (nodes) comprised in the network and which must be at least two;

– $Packets$: a finite set of all packets generated by nodes in the network;

– $STATUS$: an enumerated type that defines the different packet states (see axm1); the default Time-To-Live (TTL) parameter value assigned to packets (see axm2).

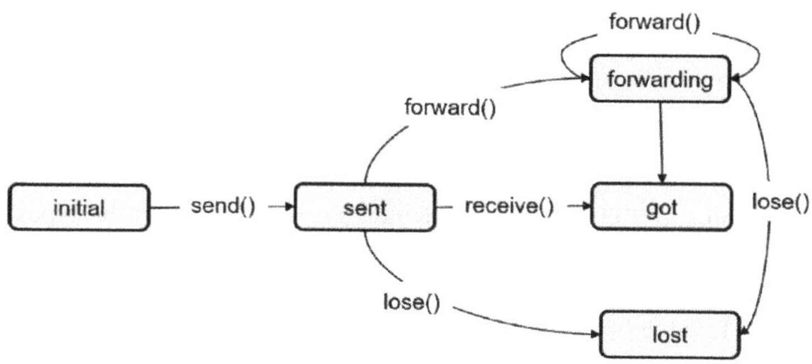

Figure 5.4. *Diagram of the operations on a packet*

CONTEXT CBL0
SETS Packets Nodes STATUS
CONSTANTS default_ttl got lost travelling sent
AXIOMS
axm1: partition(STATUS,{got},{lost},{travelling},{sent})
axm2: $default\_ttl = 255$
...
END

MACHINE CBL_m0 **SEES** CBL_c0
VARIABLES
packetStatus packetLocation packetTTL packetHops
packetSource packetDestination
INVARIANTS
inv1:packetStatus $\in$ Nodes $\times$ Packets $\twoheadrightarrow$ STATUS
...
END

Figure 5.5. *Level 1: basic functionality model of a protocol*

The machine CBL_m0 models the dynamics of the routing protocol through the communications between nodes and operations on the packets they transmit to each other or forward to each other. The machine condition is characterized by the following variables:

– $PacketStatus$ defines the current status of a packet.

– $PacketLocation$ indicates the current location of a packet.

– $PacketTTL$ contains the current TTL value of a packet.

– $PacketHops$ indicates the number of hops already made by a packet.

– $PacketSource$ identifies the sender node of a packet.

– $packetDestination$ identifies the destination node of a packet.

The machine CBL_m0 also includes events, namely, $initialization$, $sendPacket$, $receivePacket$, $forwardPacket$, $losePacket$ and $dropPacket$.

The event $sendPacket$ materializes packet transmission (Figure 5.6). The structure of the event indicates the input parameters, in this case the packet, its source, destination and TTL value. Guards make it possible to set the types of these parameters. For example, the guard $grd3$ enforces that a verification be carried out such that the TTL assigned to the packet does not exceed a threshold corresponding to the default TTL value ($default\_ttl$).

EVENT sendPacket **ANY** source destinations packet TTL
WHERE
grd1:$source \in Nodes \wedge destinations \subseteq Nodes$
grd2: $packet \in Packets$
grd3: $TTL \in \mathbb{N}_1 \wedge TTL \leq default\_ttl$
THEN
act1:$packetStatus := packetStatus \Leftarrow \{n \cdot n \in destinations | n \mapsto packet \mapsto sent\}$
act2:$packetSource := packetSource \Leftarrow \{packet \mapsto source\}$
act3:$packetDestination(packet) := destinations$
...
END

Figure 5.6. *Packet sending event*

The event $receivePacket$ models the conditions under which a packet is received by its destination node. Its input parameters are the packet, its source and its destination node. The guards verify that the initial state of this packet is the *sent* or *in transit* state. As a matter of fact, no packet can be received if it has not been previously transmitted by the source or retransmitted by a relay in the network. The event actions update the packet status as well as its location as follows:

$$packetStatus(destination \mapsto packet) := got$$

$$packetLocation(destination \mapsto packet) := destination$$

Finally, the events $forwardPacket$ and $losePacket$ model the retransmission of a packet when it is successfully carried out, or conversely its loss when the retransmission stops before the packet could reach its destination, respectively. Lost packets are removed from the transmission processes by the event $dropPacket$.

5.5.2. *Level 2: mobility and communications modeling*

This second level of abstraction formally models node dynamics as well as communications, in particular through managing the neighborhood, the positioning of nodes in road traffic and packet broadcast. It is fairly generic and as such can be utilized for modeling any protocol dedicated to mobile wireless networks. It includes a machine CBL_m1 that *sees* a context CBL_c1 (Figure 5.7). The context CBL_c1 extends the initial context CBL_c0, in the definitions proposed in section 5.4.1, namely, by introducing a finite set of hello packets ($Hello \subset Packets$). The machine CBL_m1 refines the machine CBL_m0 by introducing variables, invariants and events that can be employed for modeling communications, vehicle mobility in road traffic, as well as neighborhood discovery and management.

CONTEXT CBL_c1 **EXTENDS** CBL_c0
CONSTANTS
Hello Neighbor_expiry_time DIFF
AXIOMS
axm1: $Hello \subseteq Packets$
axm2: $Neighbor\_expiry\_time \in \mathbb{N}$
...
END

MACHINE CBL_m1 **REFINES** CBL_m0
SEES CBL_c1
VARIABLES
...
nextNodeToReceiver links neighbors positionTable receptionTime
neighboringTime time lastGotHello helloPosition up down
INVARIANTS
inv1:$links \in \mathbb{P}(Nodes \times Nodes) \wedge links \cap (Nodes \lhd id) = \varnothing$
inv2:$nextNodeToReceiver \in Packets \times Nodes \nrightarrow Nodes$
inv3:$neighbors \in Nodes \rightarrow \mathbb{P}(Nodes)$
inv4:$time \in Nodes \rightarrow \mathbb{N}$
inv5:$helloPosition \in Hello \nrightarrow \mathbb{N} \times \mathbb{N}$
...
END

Figure 5.7. *Level 2: mobility and communications modeling*

5.5.2.1. *Communications modeling*

Communications in an ad hoc network of vehicles include not only packet retransmission, but also the broadcast of control packets such as HELLOs to ensure link and neighborhood discovery and maintenance. In order to be able to formally model communications, the following variables are added to the definitions introduced in section 5.4.1:

– $nextNodeToReceiver$ identifies the next relay to which a packet in transmission will be assigned (see $inv2$, Figure 5.7). This variable is essential for the routing process of a packet from its source node to its destination node;

– $lastGotHello$ determines the last HELLO packet received at a node from another node in its vicinity. It can be defined as:

$$lastGotHello \in Nodes \times Nodes \nrightarrow Hello$$

– $time$: models the local time at each node (see $inv4$, Figure 5.7);

– $receptionTime$ refers to the time a packet was received at the destination node level. This variable is defined by the following function:

$$receptionTime \in Nodes \times Packets \nrightarrow \mathbb{N}$$

– $neighboringTime$ records the time when a node becomes a symmetric neighbor of the local node, thus allowing the management of obsolete neighbors. It is formally defined using the following function:

$$neighboringTime \in Nodes \times Nodes \nrightarrow \mathbb{N}$$

All of these variables $time$, $receptionTime$ and $neighboringTime$ are mainly used to discover and maintain neighborhoods as will be illustrated later in section 5.5.2.3. Abstract events $sendPacket$, $forwardPacket$ and $receivePacket$ are also refined at this level. Two new refined events are provided for the packet send event, namely, $sendPacket$ and $broadcast$. The first corresponds to the transmission of application packets in *unicast* mode, while the second models packet flooding, including HELLO packets. For example, Figure 5.8 shows that the abstract parameter $destinations$ vanishes from the $broadcast$ event, while new guards are added. In addition, it should be noted that the clause $with$ has been introduced, which makes it possible to emphasize the link between an abstract event and the one that refines it. This clause implies, among other things, that a packet sent in *broadcast* is likely to be received by all nodes in the network except its sender. Therefore, all occurrences of the parameter $destinations$ can be replaced by $Nodes \setminus \{source\}$. The action $act6$ is an example of this kind of substitution. The guard $grd3$ ensures that the TTL of a HELLO packet is always set to 1 when transmitting in *broadcast* in order to avoid any retransmission attempt by the nodes that receive it. The action $act7$ emphasizes that each HELLO packet sender also includes information about its position in the packet broadcast to its neighbors.

EVENT broadcast
REFINES sendPacket
ANY source packet TTL
WHERE
grd1:$source \in Nodes \wedge packet \in Packets$
grd2: $TTL \in \mathbb{N}_1 \wedge TTL \leq default\_ttl$
grd3: $packet \in Hello \Rightarrow TTL = 1$
WITH
destinations: $destinations = Nodes \setminus \{source\}$
THEN
...
act6:$packetDestination(packet) := Nodes \setminus \{source\}$
act7:$helloPosition := helloPosition \Leftarrow ((\{packet\} \cap Hello) \times \{positionTable(source)(source)\})$
END

Figure 5.8. *The event associated with packet broadcasting*

A new parameter T representing the local time at the receiving node and new guards are added in the refinement of the event associated with packet receipt. The guards clause incorporates the following three new predicates:

$$(packet \notin Hello \Rightarrow packetLocation(destination \mapsto packet) \mapsto destination \in cls(links))$$

$$T \in \mathbb{N} \wedge time(destination) \leq T$$

$$(packet \in Hello \Rightarrow destination \mapsto packet \mapsto travelling \notin packetStatus)$$

where cls is the transitive closure mathematical operator. For a relation $R \in A \leftrightarrow A$, $cls(R)$ is the transitive closure of R and admits the following general definition: $cls(R) = \bigcup_{i=0}^{\infty} R^i$, where:

$$R^i = \begin{cases} A \lhd id, & \text{if } i = 0 \\ (R \,;\, R^{i-1}), & \text{if } i \geq 1 \end{cases} \quad [5.19]$$

Actions added to the refinement of the event associated with packet receipt update the variables $time$, $receptionTime$ and $lastGotHello$. In the refinement of the packet retransmission event ($forwardPacket$), a new parameter $nextNode$ ($nextNode \in Nodes$) is introduced, along with new guards and actions. These guards are used to define the routes connecting a source node to a destination node, namely through relay nodes, while the actions essentially modify the variable $nextNodeToReceiver$ as follows:

$$nextNodeToReceiver := nextNodeToReceiver \Leftarrow \{packet \mapsto destination \mapsto nextNode\}$$

5.5.2.2. *Modeling node mobility*

The event $updatePosition$ defined in the machine CBL_m1 is used to model the motion of vehicles in the road infrastructure. As shown in Figure 5.9, this event takes the following parameters as input:

– $node$, a vehicle whose position must be updated;

– $XY \in \mathbb{N} \times \mathbb{N}$ identifies the new node position;

– $newUp$ and $newDown$ designate the upstream and downstream neighborhoods of the node, respectively.

EVENT updatePosition
ANY node XY newUp newDown
WHERE
grd1: $node \in dom(positionTable(node))$
grd2: $XY \in \mathbb{N} \times \mathbb{N}$
grd3: $newUp =$
$\{n, p \cdot n \in neighbors(node) \wedge n \mapsto p \in positionTable(node) \wedge DIFF(XY \mapsto p) > 0|n\}$
...
THEN
act1:$positionTable(node) := positionTable(node) \lhd\!\!\!- \{node \mapsto XY\}$
act2:$up(node) := newUp$
act3:$down(node) := newDown$
END

Figure 5.9. *Event updating vehicle position*

The last two parameters are automatically verified from new node positions using the guards $grd3$ and $grd4$. The event updates the variable $positionTable$ by modifying the input $positionTable(node)$ based on the data from the set $\{node \mapsto XY\}$. The actions $act2$ and $act3$ update the local node upstream ($newUp$) and downstream ($newDown$) neighbors, respectively.

5.5.2.3. *Neighborhood management modeling*

When a node receives a HELLO packet, it must update its neighbor table. In order to formally model it, an $updateNeighbor$ event is added to the machine CBL_m1 (Figure 5.10). This event consists of seven parameters, three of which ($newUp$, $newDown$ and H) are automatically computed from the previous four. The first two parameters are the source ($neighbor$) and its position XY obtained in the last HELLO packet H received from it. The parameters H (the last received HELLO packet), $newUp$ (the updated upstream neighbors) and $newDown$ (the updated downstream neighbors) are used primarily for simplicity. The other two parameters are $node$ (a destination node) and its local clock T. According to the CBL specification (Rivoirard et al. 2018), each node periodically broadcasts a HELLO packet in which it signals itself to all its direct neighbors. Therefore, when a node has not received a HELLO packet from a neighbor for a period equal to $neighbor\_expiry\_time$ since the last HELLO received from that neighbor, then that neighbor is considered to have left the neighborhood of the local node. It must therefore be removed from the neighbor table of the local node. More formally, the event $dropNeighbor$ is added to the machine CBL_m1 (Figure 5.10) in order to perform this task. The input parameters for the event $dropNeighbor$ are the node and its neighbor now assumed to be out of range. These parameters are clearly set by the guards $grd1$, $grd2$ and $grd3$. The guard $grd3$ in particular verifies the unavailability of the neighbor before applying the required actions.

EVENT updateNeighbor
ANY node neighbor T XY H newUp newDown
WHERE
grd1:$node \in Nodes \wedge neighbor \in Nodes \wedge node \neq neighbor$
grd2:$node \mapsto neighbor \in dom(lastGotHello)$
grd3:$H = lastGotHello(node \mapsto neighbor) \wedge H \in dom(helloPosition)$
grd4:$XY = helloPosition(H) \wedge node \in dom(positionTable(node))$
grd5:$T \in \mathbb{N} \wedge T \geq time(node)$
grd6:$newUp = (up(node) \setminus \{neighbor\}) \cup$
$\{p \cdot node \mapsto p \in positionTable(node) \wedge DIFF(p \mapsto XY) > 0 | neighbor\}$
grd7: $newDown = (down(node) \setminus \{neighbor\}) \cup$
$\{p \cdot node \mapsto p \in positionTable(node) \wedge DIFF(p \mapsto XY) < 0 | neighbor\}$
THEN
act1:$neighbors(node) := neighbors(node) \cup \{neighbor\}$
act2:$links := links \cup \{node \mapsto neighbor\}$
act3:$neighboringTime(node \mapsto neighbor) := T$
act4:$time(node) := T$
act5:$positionTable(node) := (\{neighbor\} \lhd positionTable(node)) \Leftarrow \{neighbor \mapsto XY\}$
act6:$up(node) := newUp$
act7:$down(node) := newDown$
END

EVENT dropNeighbor
ANY node neighbor
WHERE
grd1:$node \in Nodes$
grd2:$neighbor \in neighbors(node)$
grd3:$node \mapsto neighbor \in dom(neighboringTime)$
grd4:$\{h \cdot h \in Hello \wedge h \mapsto neighbor \in packetSource \wedge node \mapsto h \mapsto got \in packetStatusactions$
$\wedge node \mapsto h \in dom(receptionTime) \wedge receptionTime(node \mapsto h)$
$< neighboringTime(node \mapsto neighbor) + Neighbor\_expiry\_time | h\} = \varnothing$
THEN
act1:$neighbors(node) := neighbors(node) \setminus \{neighbor\}$
act2:$links := links \setminus \{node \mapsto neighbor\}$
act3:$up(node) := up(node) \setminus \{neighbor\}$
act4:$down(node) := down(node) \setminus \{neighbor\}$
act5:$positionTable(node) := \{neighbor\} \lhd positionTable(node)$
END

Figure 5.10. *Neighbor updating or deletion event*

5.5.3. *Level 3: CBL rules and properties modeling*

The third level of abstraction introduces the specific rules and properties of the case study protocol. However, the approach can be extended to the analysis of other protocols for ad hoc vehicle networks by replacing the rules and properties presented here with those specific to the targeted protocol. As shown in Figure 5.11, the model is achieved by a machine CBL_m2 that "sees" a context CBL_c2. The context CBL_c2 extends that of the second level of abstraction (CBL_c1) by adding the notion of connection time between nodes (*node connection time*, $Ctime$), initially defined in section 5.4.1. The machine CBL_m2 that *sees* the context CBL_c2 (Figure 5.11) refines the machine CBL_m1 by:

– modeling the properties of CBL;

– modeling the election of branch nodes;

– the refinement of ad hoc communications;

– modeling the management of the chain or backbone.

CONTEXT CBL_c2 **EXTENDS** CBL_c1
CONSTANTS Ctime
AXIOMS
axm1: $Ctime \in (Nodes \times Nodes) \nrightarrow \mathbb{N} \ldots$
END

MACHINE CBL_m2 **REFINES** CBL_m1
SEES CBL_c2
VARIABLES
hasType branchChoice chainUP chainDO
helloBrChoice helloChainUP helloChainDO helloSrType
...
END

Figure 5.11. *Level 3: modeling CBL rules and properties*

EVENT electBranch
ANY electing elected V
WHERE
grd1: $electing \in Nodes \wedge elected \in Nodes \wedge V \in \{0, 1\}$
grd2: $electing \mapsto elected \in dom(Ctime)$
grd3: $electing \mapsto V \in hasType(electing)$
// here guards defining NERi rules
THEN
act1: $branchChoice(electing) :=$
$branchChoice(electing) \lhd\!\!- \{e.e \in \{V\} \wedge e = 0 | electing \mapsto elected\}$
act2: $chainUP(electing) := chainUP(electing) \lhd\!\!- \{e.e \in \{V\} \wedge e = 1 | electing \mapsto elected\}$
END

Figure 5.12. *Branch node election event*

5.5.3.1. *Modeling CBL properties*

New variables are introduced in the machine CBL_m2 in order to formally model the specific rules and properties of CBL:

– $hasType$ determines the node type (branch or leaf).

– $branchChoice$ denotes the node chosen as a branch by the local node (if it is itself a leaf node).

– $chainUP$ denotes the upstream branch node chosen by the local node (if it is itself a branch node).

– $chainDO$ denotes the downstream branch node chosen by the local node (if it is itself a branch node).

The definition and role associated with each type of node has been explained in section 5.4.1. Based on the variables, constants and types introduced at this stage of the modeling, the rules and properties of CBL ($req1$, $req2$, ..., $req12$) (Table 5.1) can be formally expressed as invariants in the Event-B model of the protocol.

Properties	Corresponding invariant in the Event-B model
REQ 1	$\forall n \cdot n \in Nodes \land neighbors(n) = \varnothing \Rightarrow n \mapsto 0 \in hasType(n)$
REQ 2	$\forall n, a \cdot$ $n \in Nodes \land a \in dom(branchChoice(n)) \Rightarrow a \mapsto 0 \in hasType(n)$
REQ 3	$\forall n \cdot$ $n \in Nodes \Rightarrow (chainUP(n) \cup branchChoice(n)) \cap (Nodes \lhd id) = \varnothing$
REQ 4	$\forall n \cdot n \in Nodes \Rightarrow chainUP(n) \in Nodes \rightarrowtail\!\!\!\!\!\!\!+\;\; Nodes$
REQ 5	$\forall n \cdot n \in Nodes \Rightarrow chainDO(n) \in Nodes \rightarrowtail\!\!\!\!\!\!\!+\;\; Nodes$
REQ 6	$\forall n \cdot n \in Nodes \land n \mapsto 1 \in hasType(n) \Rightarrow$ $(chainUP(n) \cup branchChoice(n)) \rhd \{n\} \neq \varnothing$
REQ 7	$\forall a, n \cdot a \in Nodes \land a \mapsto n \in chainDO(n) \Rightarrow a \mapsto 1 \in hasType(n)$
REQ 8	$\forall a, n \cdot a \in Nodes \land a \mapsto n \in chainUP(n) \Rightarrow a \mapsto 1 \in hasType(n)$
REQ 9	$\forall a, n \cdot a \in Nodes \land a \mapsto n \in chainUP(n) \Rightarrow a \in down(n)$
REQ 10	$\forall a, n \cdot a \in Nodes \land a \mapsto n \in chainDO(n) \land n \mapsto a \in chainUP(n) \Rightarrow$ $a \in up(n)$
REQ 11	$\forall a, n \cdot a \in Nodes \land a \mapsto n \in chainDO(n) \land a \in up(n) \Rightarrow$ $n \mapsto a \in chainUP(n)$
REQ 12	$\forall n, b \cdot n \in Nodes \land n \mapsto b \in branchChoice(n) \Rightarrow b \in neighbors(n)$

Table 5.1. *Formal modeling of CBL properties in Event-B invariants*

5.5.3.2. *Modeling the election of branch nodes*

The designation of branch nodes is a crucial step in the construction of the organizational structure implemented by the CBL protocol. According to the latest proposed specification (Wahl et al. 2021), two types of elections are held simultaneously: leaf nodes elect their branch nodes, on the one hand, and branch nodes choose their upstream branch node in order to extend the chain, on the other hand. In order to formally model these processes, the $electBranch$ event is introduced (Figure 5.12). Of the three parameters associated with it, the first two, $electing$ and $elected$, designate the electing node and the elected node. The third parameter, V, is a binary variable ($V \in \{0.1\}$) intended to simplify the verification of branch node election rules ($NERi$). These rules are introduced as guards in the event $electBranch$. The first rule, $NER1$, prohibits the self-election of a node by itself. The second rule ($NER2$) requires that only nodes from the immediate neighborhood be elected. These two rules are stipulated as follows:

$$\begin{aligned}&elected \neq electing \wedge elected \in neighbors(electing) \wedge\\&\quad electing \mapsto 0 \in hasType(electing)\end{aligned}$$

Rule $NER3$ ensures that the node elected as an upstream branch is properly upstream in the neighborhood of its electing node, the latter being itself a branch node. It is transcribed as follows:

$$electing \mapsto 1 \in hasType(electing) \Rightarrow elected \in up(electing)$$

Rule $NER4$ is intended to foster chain uniqueness by avoiding the creation of parallel chains. It formally achieves this objective by requiring branch nodes to have no more than one upstream branch:

$$V = 1 \Rightarrow chainUP(electing) \rhd up(electing) = \varnothing$$

Rule $NER5$ requires any leaf node in which none of the neighbors has branch node status to elect the one with which it achieves the highest connection time:

$$\begin{aligned}&elected \mapsto 0 \in hasType(electing) \wedge V = 0 \Rightarrow\\&\quad \neg(\exists n \cdot n \in neighbors(electing) \wedge electing \mapsto n \in dom(Ctime) \wedge\\&\quad Ctime(electing \mapsto elected) < Ctime(electing \mapsto n))\end{aligned}$$

The $NER6$ rule specifies a similar rule regarding the election of the upstream branch node. The branch node that has no upstream neighbors with branch node status elects the neighbor with the longest connection time as its upstream branch node. Formally, this rule is written as follows:

$$\begin{aligned}&\{elected \mapsto 0\} \subseteq hasType(electing) \wedge V = 1 \Rightarrow\\&\quad \neg(\exists n \cdot n \in up(electing) \wedge electing \mapsto n \in dom(Ctime) \wedge\\&\quad Ctime(electing \mapsto elected) < Ctime(electing \mapsto n))\end{aligned}$$

5.5.3.3. *Refining communication between nodes*

The CBL protocol relies on the exchange of HELLO packets between the network nodes to perform in a distributed manner the operations that lead to the formation of clusters and the backbone (chain) that interconnects them. For the protocol to be effective, the HELLO packets exchanged must contain information useful to these distributed operations. These are particularly the sender of the HELLO packet, its position in the road infrastructure, its type (leaf node or branch node), if necessary its choices of branch node, upstream branch node and downstream branch node, as well as a list of its direct neighbors. To formally model this information, the following variables are introduced:

– $helloBrChoice$ denotes the branch node chosen by the sender of the HELLO packet (if it is itself a leaf node).

– $helloChainUP$ represents the upstream branch node chosen by the sender of the HELLO packet (if it is itself a branch node).

– $helloChainDO$ represents the downstream branch node chosen by the sender of the HELLO packet (if it is itself a branch node).

– $helloSrType$ indicates the type of the sender of the HELLO packet (leaf node or branch node).

These variables are introduced into the model as partial functions defined as part of the invariants of the machine CBL_m2, according to their initial definitions in section 5.4.1. The event associated with packet delivery is refined in order to incorporate information from HELLO packets. The following guard, relative to the property $REQ12$ of CBL, is added to force leaf nodes with neighbors to choose a branch node before broadcasting their HELLO message:

$$\begin{array}{l} neighbors(source) \neq \varnothing \wedge source \mapsto 0 \in hasType(source) \Rightarrow \\ \qquad 1 \in hasType(source)[neighbors(source)] \end{array}$$

New substitution actions are also introduced in the refined version of the $broadcast$ event to update the variables *helloBrChoice*, *helloChainUP*, *helloChainDO* and *helloSrType* as follows:

$$\begin{array}{l} helloBrChoice := helloBrChoice \Leftarrow ((\{packet\} \cap Hello) \times ran(\{source\} \\ \qquad \lhd branchChoice(source))) \\ helloChainUP := helloChainUP \Leftarrow ((\{packet\} \cap Hello) \times ran(\{source\} \\ \qquad \lhd chainUP(source))) \\ helloSrType := helloSrType \Leftarrow ((\{packet\} \cap Hello) \\ \qquad \times \{hasType(source)(source)\}) \\ helloChainDO := helloChainDO \Leftarrow ((\{packet\} \cap Hello) \times ran(\{source\} \\ \qquad \lhd chainDO(source))) \end{array}$$

5.5.3.4. *Modeling chain updates*

Several events trigger actions at each node to update the chain formed by CBL. These events include changes in the neighborhood due to the loss of connections with certain nodes, the overtake of a branch node by its upstream branch in road traffic, etc. Broadly speaking, these events can be restricted to three main groups, namely, link loss, changes in node positions and the reception of HELLO packets informing of changes in the neighborhood (for instance, the election of a branch node or of an upstream or downstream branch, etc.).

In order to formally model these events, refinements were made to the abstract events $updatePosition$, $dropNeighbor$ and $updateNeighbor$. In addition, two new events have been added, namely, $turnIntoBranch$ and $turnIntoLeaf$, whose function is to manage the change in node status according to the rules $TCRi$. The event $updateNeighbor$ is refined in the following way:

$$\begin{array}{l} hasType(node) := newTypes \\ chainDO(node) := newChainDO \\ chainUP(node) := newChainUP \\ branchChoice(node) := newBrChoice \end{array}$$

where $newTypes$, $newChainDO$, $newChainUP$ and $newBrChoice$ refer to the parameters introduced in the refinement of the event, and constitute the new values of the functions $hasType(node)$, $chainDO(node)$, $chainUP(node)$ and $branchChoice(node)$ after adding or updating information about neighbors (status, branch choice, etc.). For example, the values of the parameters $newChainUP$ and $newChainDO$ are computed as follows:

$$\begin{array}{l} newChainUP = chainUP(node) \Leftarrow (\{neighbor\} \times ran(\{H\} \\ \qquad\qquad\qquad\qquad\qquad\qquad \lhd helloChainUP)) \\ newChainDO = \{x \mapsto y | y \mapsto x \in newChainUP\} \end{array}$$

The other parameters are calculated in the same way. The refinement of the event $dropNeighbor$ makes it possible to locally update the chain when the link to a neighbor is lost. Compared to the abstract event, this refinement introduces the following actions:

$$\begin{array}{l} branchChoice(node) := branchChoice(node) \setminus \{node \mapsto neighbor\} \\ chainUP(node) := chainUP(node) \setminus \{node \mapsto neighbor\} \\ chainDO(node) := chainDO(node) \setminus \{node \mapsto neighbor\} \\ hasType(node) := \{neighbor\} \lhd hasType(node) \end{array}$$

Updating the position of node Ni requires us to verify whether Ni has overtaken node Nj that would have elected it as the upstream branch, in which case the link between Ni and Nj in the chain should be removed by an action added to the event $updatePosition$:

$$chainUP(node) := newChainUP$$

where $newChainUP$ is a new parameter denoting the new upstream branch, which is automatically computed as follows:

$$(down(node) \setminus newDown) \lhd chainUP(node) \qquad [5.20]$$

For this, the following conditions, expressed as guards, must be met:

$$(\forall a \cdot a \mapsto node \in chainDO(node) \wedge node \mapsto a \in chainUP(node) \Rightarrow a \in newUp)$$
$$\wedge(\forall a \cdot a \mapsto node \in chainDO(node) \wedge a \in newUp \Rightarrow node \mapsto a \in newChainUP)$$
$$\wedge(newChainUP \in Nodes \twoheadrightarrow\!\!\!\!\!\!\mapsto Nodes) \wedge (\forall n \cdot n \in Nodes \wedge n \mapsto 1 \in hasType(n)$$
$$\Rightarrow (newChainUP \cup branchChoice(node)) \rhd \{n\} \neq \varnothing)$$

EVENT turnIntoBranch **ANY** leaf electing
WHERE
grd1:$(branchChoice(leaf) \cup chainUP(leaf))$
grd2:$leaf \mapsto 0 \in hasType(leaf)$
grd3:$neighbors(leaf) \neq \varnothing$
grd4:$leaf \notin dom(branchChoice(leaf))$
grd5:$electing \mapsto leaf \in chainUP(leaf) \Rightarrow chainDO(leaf) \rhd down(leaf) = \varnothing$
THEN
act1:$hasType(leaf) := hasType(leaf) \Leftarrow \{leaf \mapsto 1\}$
act2:$chainDO(leaf) :=$
$\quad chainDO(leaf) \Leftarrow \{x \mapsto y | x \mapsto y \in \{electing \mapsto leaf \wedge y \mapsto x \in chainUP(leaf)\}$
END
EVENT turnIntoLeaf **ANY** node
WHERE
grd1:$\{node\} \lhd chainDO(node) \neq \varnothing \wedge chainUP(node) \rhd \{node\} = \varnothing$
grd2:$node \mapsto 1 \in hasType(node)$
grd3:$finite(ran(\{node\} \lhd chainDO(node)) \cap down(node))$
grd4:$card(ran(\{node\} \lhd chainDO(node)) \cap down(node)) \leq 1$
THEN
act1:$hasType(node) := hasType(node) \Leftarrow \{node \mapsto 0\}$
act2:$chainDO(node) := \{node\} \lhd chainDO(node)$
END

Figure 5.13. *Node status update event*

The $turnIntoBranch$ event is used to manage the change in status from leaf node to branch node once it has been elected by at least one of its neighbors (see $grd1$ in Figure 5.13). This change in status occurs after receiving a HELLO packet from a neighbor. The guard $grd3$ verifies the property $REQ6$ of CBL that forbids self-election. The event $turnIntoLeaf$ is used to change the status of a branch node to leaf node when the latter overtakes the branch node that previously elected it as an upstream branch (see Figure 5.13). The guard $grd1$ sets a precondition on this change of status following rule $TCR3$.

5.6. Model validation

The method adopted during model validation is based on a two-phase approach: model validation by model animation using the `ProB` model verification tool and proof of model correction by unloading POs.

5.6.1. *Model validation by animation*

The animation tool `ProB` (Leuschel 2003) allows for both requirement validation and error detection. It also can be used to correct the latter before initiating the proof phase, which is generally long and complex. In practice, animation cannot be achieved with an abstract model and requires a concrete model. To this end, the machine CBL_m3 that extends the machine CBL_m2 is introduced at the same time as the context CBL_c3 that it *sees*. Unlike the machine, which does not introduce any new events or variables, the context defines constants and axioms that express (Figure 5.14) the sets and functions defined in the CBL Event-B model.

CONTEXT CBL_c3 **EXTENDS** CBL_c2
CONSTANTS
v1 v2 v3 v4 v5 v6
hv1 hv2 hv3 hv4 hv5 hv6 msg
AXIOMS
axm1: $partition(Nodes, \{v1\}, \{v2\}, \{v3\}, \{v4\}, \{v5\}, \{v6\})$
axm2: $partition(Hello, \{hv1\}, \{hv2\}, \{hv3\}, \{hv4\}, \{hv5\}, \{hv6\})$
axm3: $partition(Traffics, \{1\}, \{2\}, \{3\})$
axm4: $partition(position, \{v1 \mapsto 1 \mapsto 2\}, \{v2 \mapsto 2 \mapsto 5\}, \{v3 \mapsto 3 \mapsto 4\}, \{v4 \mapsto 2 \mapsto 10\}, \{v5 \mapsto 1 \mapsto 10\}, \{v6 \mapsto 3 \mapsto 15\})$
axm6: $partition(Packets, \{msg\}, \{hv1\}, \{hv2\}, \{hv3\}, \{hv4\}, \{hv5\}, \{hv6\})$
axm7: $partition(timeConnection,$
$\{v1 \mapsto v2 \mapsto 12\}, \{v1 \mapsto v3 \mapsto 12\}, \{v1 \mapsto v4 \mapsto 12\}, \{v1 \mapsto v5 \mapsto 12\}, \{v1 \mapsto v6 \mapsto 12\},$
$\{v2 \mapsto v1 \mapsto 32\}, \{v2 \mapsto v3 \mapsto 12\}, \{v2 \mapsto v4 \mapsto 12\}, \{v2 \mapsto v5 \mapsto 12\}, \{v2 \mapsto v6 \mapsto 12\},$
$\{v3 \mapsto v1 \mapsto 2\}, \{v3 \mapsto v2 \mapsto 12\}, \{v3 \mapsto v4 \mapsto 6\}, \{v3 \mapsto v5 \mapsto 12\}, \{v3 \mapsto v6 \mapsto 12\},$
$\{v4 \mapsto v1 \mapsto 42\}, \{v4 \mapsto v2 \mapsto 32\}, \{v4 \mapsto v3 \mapsto 8\}, \{v4 \mapsto v5 \mapsto 12\}, \{v4 \mapsto v6 \mapsto 12\},$
$\{v5 \mapsto v1 \mapsto 52\}, \{v5 \mapsto v2 \mapsto 22\}, \{v5 \mapsto v3 \mapsto 23\}, \{v5 \mapsto v4 \mapsto 12\}, \{v5 \mapsto v6 \mapsto 12\},$
$\{v6 \mapsto v1 \mapsto 12\}, \{v6 \mapsto v2 \mapsto 4\}, \{v6 \mapsto v3 \mapsto 52\}, \{v6 \mapsto v4 \mapsto 12\}, \{v6 \mapsto v5 \mapsto 12\})$
END

Figure 5.14. *CBL model implementation for animation*

Figure 5.15 describes the animation window of the ProB tool, which consists of three parts:

– Part (1) describes event triggering and the verification of constraints. It is usually on the left side of the tool window.

– Part (2) presents the model status.

– Part (3) reports invariant violations and specification errors.

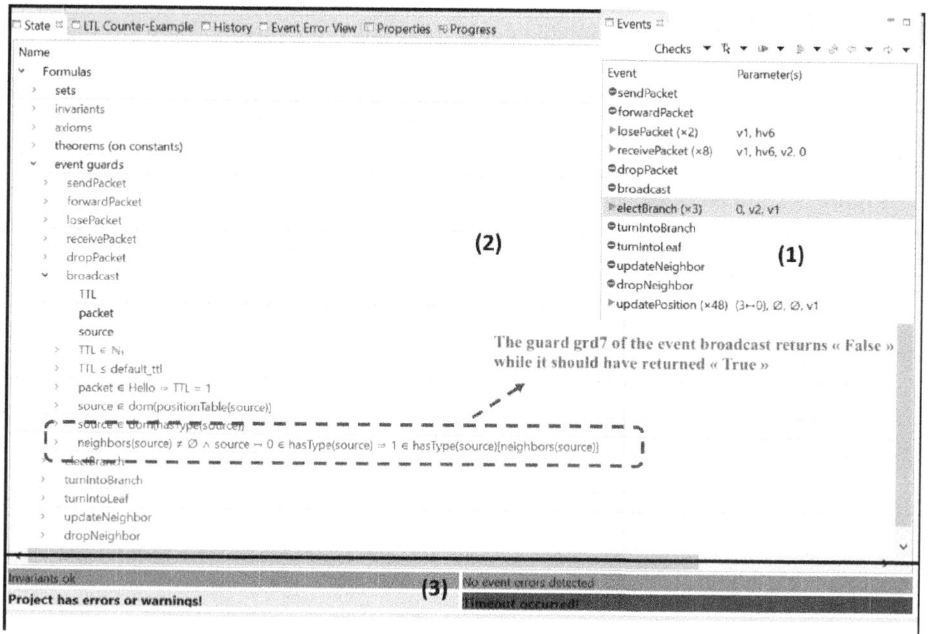

Figure 5.15. *CBL model animation ProB window. For a color version of this figure, see www.iste.co.uk/wahl/design.zip*

When the animation context is launched, only the initialization event is activated. The *updatePosition* and *broadcast* events are then successfully triggered. After the broadcasting of a HELLO packet by a node, the events *receivePacket* and *losePacket* are enabled. The event *updateNeighbor* is triggered several times following the event *receivePacket*. Regarding the sequence of the animation scenario, Figure 5.15 shows that the events *broadcast* and *turnIntoBranch* are disabled after executing the *updateNeighbor* and *electBranch* events. However, it can be observed that no node is transformed into a branch node and therefore the chain is not formed. This situation suggests that some nodes were unable to broadcast the HELLO message noticing their choice of branch node to their neighborhood. It seems that this comes from guard *grd*7 (Figure 5.15), which verifies that at least one of the neighbors of a leaf node is a branch node. A correction is achieved by modifying the guard as follows:

$$neighbors(source) \neq \varnothing \land source \mapsto 0 \in hasType(source)$$
$$\Rightarrow \{source\} \lhd branchChoice(source) \neq \varnothing$$

The counterexamples in the ProB tool can serve as guides for correcting models and detecting specification errors that could have failed the prover. These corrections may as well concern invariant violations and modifications to guards or actions. The proposed CBL model was thus validated based on the scenarios representing the situations that can occur in an ad hoc network of vehicles in a road environment.

Figure 5.16 depicts the result of an Event-B animation sequence, the details of which are shown in Table 5.2.

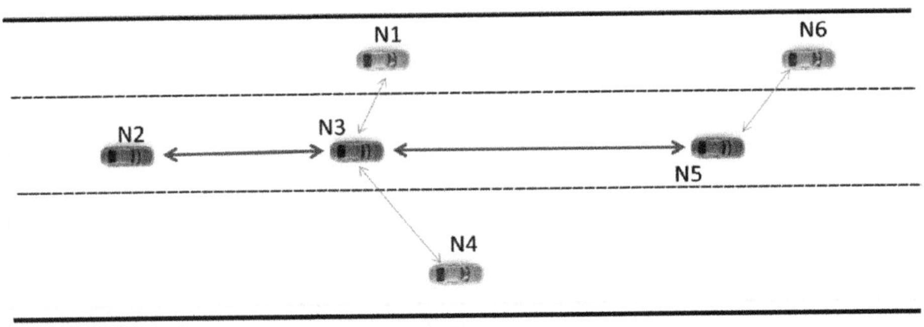

Figure 5.16. *Example of a chain resulting from an Event-B animation sequence. For a color version of this figure, see www.iste.co.uk/wahl/design.zip*

5.6.2. *Model correction by offloading proof obligations*

In order to verify the model during the design process, its POs are offloaded so as to ensure that:

– the model initialization leads to a state where the invariant is valid;

– when a machine in the model is in a state in which the invariant is valid, any active event must lead to a state that preserves that validity;

– concrete events can only be triggered under the same circumstances as abstract events;

– the occurrence of any concrete event implies the occurrence of the associated abstract event such that the state satisfies all relative invariants.

Triggered event	Status of machine variables
$1 - broadcast(N1, h11)$	$packetStatus = \{N1 \mapsto h11 \mapsto sent\}$ $packetSource = \{h11 \mapsto N1\}, ...$
$2 - broadcast(N3, h31)$	$packetStatus = \{N1 \mapsto h11 \mapsto sent,$ $N3 \mapsto h31 \mapsto sent\}$ $packetSource = \{h11 \mapsto N1, h31 \mapsto N3\},$ $...$
$3 - receivePacket(N1, h31, N3)$	$packetStatus = \{N1 \mapsto h11 \mapsto sent,$ $N3 \mapsto h31 \mapsto sent, N3 \mapsto h11 \mapsto got\}...$
$4 - updateNeighbor(h31, 0, N1, ...)$	$Neighbors = \{N1 \mapsto N3, ..\},$ $up = \{N1 \mapsto N3\}, down = \{N1 \mapsto \{\}\}, ...$
$5 - receivePacket(N2, h31, N3)$	$packetStatus = \{N1 \mapsto h11 \mapsto sent,$
$6 - updateNeighbor(h31, N2, ...)$	$Neighbors = \{N1 \mapsto \{N3\}, N2 \mapsto \{N3\}, ...\},$ $up = \{N1 \mapsto \{N3\}\},$ $down = \{N1 \mapsto \{\}, N2 \mapsto \{N3\}\}, ...$
$7 - receivePacket(N4, h31, N3)$	$packetStatus = \{N1 \mapsto h11 \mapsto sent,$ $N3 \mapsto h31 \mapsto sent, N3 \mapsto h11 \mapsto got,$ $N2 \mapsto h31 \mapsto got, N4 \mapsto h31 \mapsto got\}, ...$
$8 - updateNeighbor(h31, N4, ...)$	$Neighbors = \{N1 \mapsto \{N3\}, N2 \mapsto \{N3\},$ $N4 \mapsto \{N3\}...\},$ $up = \{N1 \mapsto \{N3\}, N4 \mapsto \{N3\}\},$ $down = \{N1 \mapsto \{\}, N2 \mapsto \{N3\}\}, ...$
$9 - electBranch(0, N3, N1)$	$branchChoice = \{N1 \mapsto \{N1 \mapsto N3\}\},$ $chainUP = \{N1 \mapsto \{\}\}, ...$
$10 - broadcast(N1, h12)$	$packetStatus = \{N1 \mapsto h12 \mapsto sent,$ $N1 \mapsto h11 \mapsto sent, N3 \mapsto h31 \mapsto sent,$ $N3 \mapsto h11 \mapsto got, N2 \mapsto h31 \mapsto got,$ $N4 \mapsto h31 \mapsto got\}, ...$ $packetSource = \{h11 \mapsto N1,$ $h31 \mapsto N3, h12 \mapsto N1\}, ...$
$11 - receivePacket(N3, h12, N1)$	$packetStatus = \{N1 \mapsto h12 \mapsto sent,$ $N1 \mapsto h11 \mapsto sent, N3 \mapsto h31 \mapsto sent,$ $N3 \mapsto h11 \mapsto got, N2 \mapsto h31 \mapsto got,$ $N4 \mapsto h31 \mapsto got, N3 \mapsto h12 \mapsto got\}, ...$
$12 - updateNeighbor(h12, N3, ...)$	$Neighbors = \{N1 \mapsto \{N3\}, N2 \mapsto \{N3\},$ $N4 \mapsto \{N3\}, N3 \mapsto \{N1\}...\},$ $up = \{N1 \mapsto \{N3\},$ $N4 \mapsto \{N3\}, N3 \mapsto \{\}\},$ $down = \{N1 \mapsto \{\}, N2 \mapsto \{N3\},$ $N3 \mapsto \{N1\}\}, ...$

Triggered event	Status of machine variables
13 – $turnIntoBranch(N3, N1)$	$hasType = \{N3 \mapsto \{N3 \mapsto 1, N1 \mapsto 0\}, ...\}$, $chainDO = \{N3 \mapsto \{\}, ..\}, ...$
14 – $broadcast(N3, h32)$	$packetStatus = \{N3 \mapsto h32 \mapsto sent, ...\}, ...$ $packetSource = \{h32 \mapsto N3, ..\}, ...$
15 – $receivePacket(N2, h32, N3)$	$packetStatus = \{N2 \mapsto h32 \mapsto got,$ $N3 \mapsto h32 \mapsto sent, ...\}, ...$
16 – $updateNeighbor(h32, N2, ...)$	$Neighbors = \{N1 \mapsto \{N3\}, N2 \mapsto \{N3\},$ $N4 \mapsto \{N3\}, N3 \mapsto \{N1\}...\}$, $up = \{N1 \mapsto \{N3\}, N4 \mapsto \{N3\}, N3 \mapsto \{\}\}$, $down = \{N1 \mapsto \{\}, N2 \mapsto \{N3\},$ $N3 \mapsto \{N1\}\}, ...$
17 – $electBranch(0, N3, N2)$	$branchChoice = \{N1 \mapsto \{N1 \mapsto N3\},$ $N3 \mapsto \{N2 \mapsto N3\}, ...\}$, $chainUP = \{N1 \mapsto \{\}\}, ...$
18 – $receivePacket(N4, h32, N3)$	$packetStatus = \{N3 \mapsto h32 \mapsto sent,$ $N4 \mapsto h32 \mapsto got, ...\}, ...$
19 – $updateNeighbor(h32, N4, ...)$	$Neighbors = \{N4 \mapsto \{N3\}, ...\}$, $up = \{N4 \mapsto \{N3\}, N4 \mapsto \{N3\}, ...\}$, $down = \{N4 \mapsto \{\}, ...\}, ...$
20 – $electBranch(0, N3, N4)$	$branchChoice = \{N1 \mapsto \{N1 \mapsto N3\},$ $N3 \mapsto \{N2 \mapsto N3\}, N4 \mapsto \{N4 \mapsto N3\}, ...\}$, $chainUP = \{N1 \mapsto \{\}, ...\}, ...$
21 – $broadcast(N5, h51)$	$packetStatus = \{N5 \mapsto h51 \mapsto sent, ...\}, ...$ $packetSource = \{h51 \mapsto N5, ...\}, ...$
22 – $receivePacket(N6, h51, N5)$	$packetStatus = \{N6 \mapsto h51 \mapsto got,$ $N5 \mapsto h51 \mapsto sent, ...\}, ...$
23 – $updateNeighbor(h51, N6, ...)$	$Neighbors = \{N6 \mapsto \{N5\}, ...\}$, $up = \{N6 \mapsto \{N5\}, ...\}...$
24 – $turnIntoBranch(N5, N6)$	$hasType = \{N5 \mapsto \{N5 \mapsto 1, N6 \mapsto 0\}, ...\}, ...$
25 – $broadcast(N3, h33)$	$packetStatus = \{N3 \mapsto h33 \mapsto sent, ...\}, ...$ $packetSource = \{h33 \mapsto N3, ...\}, ...$
26 – $receivePacket(N5, h33, N3)$	$packetStatus = \{N5 \mapsto h33 \mapsto got,$ $N3 \mapsto h33 \mapsto sent, ...\}, ...$
27 – $updateNeighbor(h33, N5, ...)$	$Neighbors = \{N5 \mapsto \{N3, N6\}, ...\}$, $up = \{N5 \mapsto \{N3\}, ...\}$, $down = \{N5 \mapsto \{N6\}, ...\}, ...$
28 – $electBranch(1, N3, N5)$	$branchChoice = \{N5 \mapsto \{N6 \mapsto N5, ...\}, ...\}$, $chainUP = \{N5 \mapsto \{N5 \mapsto N3\}\}, ...$

Table 5.2. *Animation sequence to verify the CBL Event-B model*

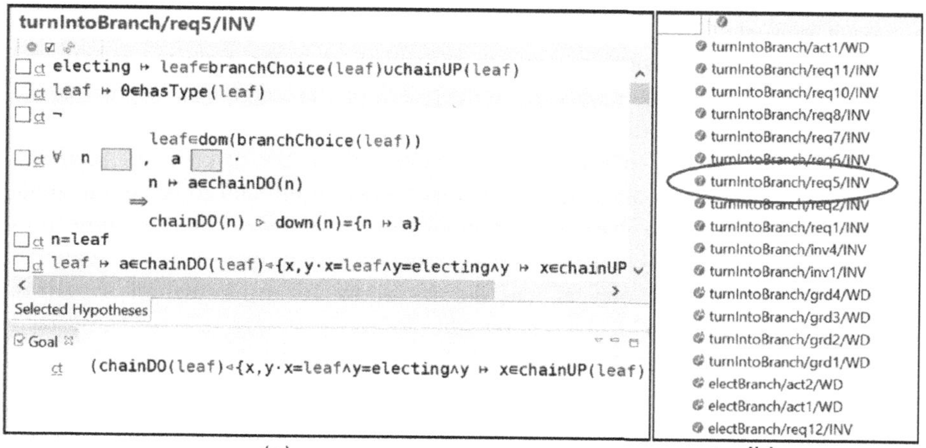

(a) (b)

Figure 5.17. *Examples of proof obligations of the CBL model in Rodin. For a color version of this figure, see www.iste.co.uk/wahl/design.zip*

POs refer to properties that can be proven in an Event-B model. Figure 5.17 shows examples of POs. Those that are offloaded are indicated with the symbol (shown in part b of the figure, with a green check mark and the letter A), while those that are not can be recognized by the symbol (circled in part b of the figure). POs that are automatically offloaded also display the letter "A". When the prover fails in automatically offloading a PO, an expert may step in to carry out an interactive proof. Figure 5.18 shows the distribution of POs that were automatically offloaded in contrast to those that were manually offloaded through interactive proof. For example, the use of the cardinal operator implies that the operand is a finite set, which makes it more difficult to offload the POs. Similarly, universal ($\forall$) and existential ($\exists$) quantifiers can be difficult when instantiating quantified assumptions. In the proposed CBL model, an example of this kind of PO is $turnIntoBranch/req5/INV$, meaning that the event $turnIntoBranch$ preserves the invariant $req5$ of the machine CBL_m2. Figure 5.17(a) shows that this PO is not automatically proven due to a lack of assumptions. The issue was fixed by adding a new guard to the $turnIntoBranch$ event (Figure 5.13):

$$electing \mapsto leaf \in chainUP(leaf) \Rightarrow chainDO(leaf) \rhd down(leaf) = \varnothing$$

5.7. Conclusion

An example of another approach to the design and evaluation of a routing protocol has been presented in this chapter. Conversely to the functional design approach presented in Chapter 2, which consists of continuously orientating the actions of the protocol with the hope that it will lead to the expected results, the formal approach developed in this chapter consists, on the contrary, of expressing the expected protocol properties that will be verified under the defined conditions regardless of the actions of the protocol. Nevertheless, the diversity of scenarios introduced in the context of vehicular networks makes it difficult to be certain about the completeness of the basic assumptions (axioms) underlying the properties expressed. Animation tools integrated into formal modeling tools can only support a very limited number of objects and do not allow for exploring all scenarios, especially those that would make it possible to verify scaling results. These facts open up new research avenues that we will develop in Chapter 6.

Proof Control | Statistics | Rodin Problems

Element Name	Tot.	Aut.	Man.	Rev.	Und.
CCM4CBL	**281**	**174**	**107**	**0**	**0**
CBL_c0	1	1	0	0	0
CBL_c1	3	0	3	0	0
CBL_c2	0	0	0	0	0
CBL_m0	30	27	3	0	0
CBL_m1	96	66	30	0	0
CBL_m2	151	80	71	0	0

Figure 5.18. *Statistic synthesis of the proofs of the CBL model in Rodin*

5.8. References

Abrial, J.-R. (2005). *The B-Book: Assigning Programs to Meanings*. Cambridge University Press, Cambridge.

Abrial, J.-R. (2010). *Modeling in Event-B: System and Software Engineering*. Cambridge University Press, Cambridge.

Barner, J. (2005). A lightweight formal method for the prediction of non-functional system properties. PhD Thesis, Université Friedrich Alexander, Erlangen.

Bowen, J.P. and Hinchey, M.G. (2006). Ten commandments of formal methods... ten years later. *Computer*, 39(1), 40–48. doi: 10.1109/MC.2006.35.

Cavalli, A. (2001). *Ingénierie des protocoles et qualité de service*. Hermes-Lavoisier, Paris.

Chebbi, E. (2019). De la modélisation formelle à la simulation à évènements discrets : application à la conception et à l'évaluation de protocoles sûrs et sécurisés pour les communications dans les transports. PhD Thesis, Université du Littoral Côte d'Opal, Dunkirk [Online]. Available at: http://www.theses.fr/2019DUNK0538.

Chebbi, E., Sondi, P., Ramat, E. (2019). A formal model for the chain-branch-leaf clustering scheme in OLSR based vehicular ad hoc networks using Event-B. *Procedia Computer Science*, 151, 935–940.

Diaz, J., Arroyo, D., Rodriguez, F.B. (2014). A formal methodology for integral security design and verification of network protocols. *Journal of Systems and Software*, 89, 87–98. doi: 10.1016/j.jss.2013.09.020.

Fakhfakh, F., Tounsi, M., Mosbah, M. (2019). An evaluative review of the formal verification for vanet protocols. In *2019 15th International Wireless Communications Mobile Computing Conference (IWCMC)*. IEEE, Tangier. doi: 10.1109/ IWCMC.2019.8766783.

Fu, C. and Zheng, K. (2019). Formal modeling and analysis of ad hoc zone routing protocol in Event-B. *International Journal on Software Tools for Technology Transfer*, 21(2), 165–181.

Jackson, D. (ed.). (2001). Lightweight formal methods. In *International Symposium of Formal Methods Europe*. Springer, Berlin.

Kamali, M. and Petre, L. (2016). Modelling link state routing in Event-B. In *2016 21st International Conference on Engineering of Complex Computer Systems (ICECCS)*. IEEE, Dubai.

Leduc, G. and Kuty, L. (2001). Spécification formelle des mécanismes de support des qualités de service dans l'internet. In *Ingénierie des protocoles et qualité de service*, Cavalli, A. (ed.). Hermes-Lavoisier, Paris.

Lee, G. (2014). How to formally model features of network security protocols. *International Journal of Security and Its Applications*, 8(1), 423–432.

Leuschel, M.B.M. (2003). ProB: A model checker for B. In *FME 2003: Formal Methods, LNCS 2805*. Springer, Berlin.

Méry, D. and Singh, N.K. (2010). EB2C: A tool for Event-B to C conversion support. In *8th IEEE International Conference on Software Engineering and Formal Methods (SEFM)* [Online]. Available at: https://hal.inria.fr/inria-00540006.

Méry, D. and Singh, N.K. (2011). Analysis of DSR protocol in Event-B. In *Stabilization, Safety, and Security of Distributed Systems*, Défago, X., Petit, F., Villain, V. (eds). Springer, Berlin, Heidelberg. doi: 10.1007/978-3-642-24550-3_30.

Rivoirard, L., Wahl, M., Sondi, P., Berbineau, M., Gruyer, D. (2018). Chain-branch-leaf: A clustering scheme for vehicular networks using only V2V communications. *Ad Hoc Networks*, 68, 70–84. doi: 10.1016/j.adhoc.2017.10.007.

Rushby, J. (1993). Formal methods and the certification of critical systems. Technical Report, SRI-CSL-93-7, Computer Science Laboratory, SRI International, Menlo Park [Online]. Available at: http://www.csl.sri.com/papers/csl-93-7/.

Singh, A. and Singh, V. (2013). Formal modeling of distance vector routing protocol using Event-B. *Electronic and Electric Engineering*, 3, 91–98.

Sondi, P., Abbassi, I., Ramat, E., Chebbi, E., Graiet, M. (2021). Modeling and verifying clustering properties in a vehicular ad hoc network protocol with Event-B. *Scientific Reports*, 11(17620), 1–14. doi: 10.1038/s41598-021-97063-3.

Wahl, M., Sondi, P., Rivoirard, L. (2021). Enhanced CBL clustering performance versus GRP, OLSR and AODV in vehicular ad hoc networks. *Telecommunication Systems*, 76, 525–540. doi: 10.1007/s11235-020-00734-1.

Zowghi, D. and Gervasi, V. (2004). Erratum to on the interplay between consistency, completeness, and correctness in requirements evolution. *Information and Software Technology*, 46(11), 763–779. doi: 10.1016/j.procs.2018.04.049.

6

Integration of Ad Hoc Components in a Global Infrastructure

6.1. Introduction

In Chapter 1, we mentioned the telecommunication network for the railway sector as the first full-scale experiment on a scale comparable to a regional road network. On this occasion, we stressed the need, unlike the rail sector, which has opted for a dedicated infrastructure, to steer the choice of future communication infrastructures for land transport toward a shared solution with other consumer services in order to better support development and reduce costs.

Another important aspect mentioned in Chapter 1 is the complementarity that could be implemented between self-organizing components through ad hoc communications with a communications network infrastructure in areas where it is available and is operational. We have actually pointed out that any traffic that is routed by ad hoc communications would make it possible to preserve the equivalent capabilities at the infrastructure level, which would thus be available to process additional traffic.

In this chapter, we address some key points, in the light of the work that has been presented throughout this book, to open up perspectives related to the design and evaluation of this future infrastructure for communications in land transport.

6.2. Ensuring connectivity and services

Driver assistance systems applied to the automotive sector enable safety (with risk minimization), mobility (traffic optimization and congestion minimization) and environmental (energy consumption optimization and pollution minimization) problems to be dealt with. These in-vehicle embedded systems and applications (so-called Advanced Driver Assistance Systems (ADAS)) rely on the use of optimized communication, localization and perception functions of the road environment.

Studies in the scientific literature, mentioned in particular in Chapter 4, have brought forward the interest of cooperative localization or ego-localization in a constellation of road vehicles for driver assistance applications. The idea is for each communicating host vehicle to use the information received from neighboring vehicles for refining and improving their own localization. For these wireless multi-sensor systems, communications involving the infrastructure alone (V2I) could prove unsuitable in certain contexts, or even ineffective in the presence of occasional degraded conditions. While the advantages of a centralized approach are accuracy and capacity, it requires an infrastructure equipped with radio access points at regular intervals to be available. Yet, it is clearly visible that this could not be guaranteed, mainly for reasons related to return on investment, especially in road sections that are generally less frequented. In addition, the information transmitted through the infrastructure can present an increased risk of obsolescence compared to direct vehicle-to-vehicle (V2V) communication when the vehicles involved are in close proximity.

The implementation of specific distributed application functions through ad hoc networks would contribute to the overall reduction of the load on the infrastructure (compute time, bandwidth, etc.), which would allow the processing of more traffic and provide applications with greater modularity and flexibility. However, distributed media access and routing algorithms in ad hoc networks are more complex and their energy loads add up to those of the information transmission functions specific to this type of network. It is therefore necessary to develop routing protocols capable of adapting to this very specific context. In Chapter 2, we have shown, through the case study protocol, the many and diverse aspects that should be taken into account for the design of such a protocol. The proposed distributed and collaborative *Chain-Branch-Leaf* (CBL) *clustering* algorithm, when a communication infrastructure is lacking or failing, provides the application layer of vehicles operating in the area with the services of a dynamic and stable virtual infrastructure.

This virtual infrastructure connects all vehicles through a backbone (chain) formed by elected nodes, the branch nodes, which act as access points. All nonelected nodes, leaf nodes, when they move in a portion of the road covered by a branch node, choose the latter as a relay for their traffic. The CBL chain structure allows for the introduction of the targeted communication services. First of all, it is possible to define the directions in which messages are broadcast (within the group, upstream of the traffic, downstream and to the whole VANET) in order to confine the impact of this broadcast. Second, it is compatible with the implementation of the four communication modes specified by the *GeoNetworking* standard (point-to-point, point-to-multipoint, *GeoAnyCast, GeoBroadcast*). The evaluations proposed in Chapters 2 and 4 show that this virtual infrastructure limits the impact of routing traffic to give priority to application traffic, because CBL only requires the transmission of periodic HELLO messages for neighborhood discovery and distributed network structuring. Standardized messages such as CAM (*Cooperative Awareness Message*) and DENM (*Decentralized Environmental Notification Message*) could be used for this purpose, which would possibly eliminate the need for additional HELLO messages. In terms of application traffic, CBL enables, on the one hand, close communications necessary for the exchange of variables and messages between neighboring driven or autonomous vehicles, and, on the other hand, multi-hop communications typically necessary for the exchange of alert notification messages and more broadly for road network management.

The analysis of the structural results of CBL-OLSR clearly shows the effectiveness of this protocol for vehicular networks. In scenarios with high vehicle density, only 20% of the nodes in the network are branch nodes relaying traffic. For a crossing of a motorway section over an average period of 150 s, the nodes remain in the branch state for more than 50 s on average (i.e. one-third of the time), while leaf-branch links last for an average of 35 s (i.e. a quarter of the time). This closeness of the duration of the branch nodes (cluster heads) to the duration of their association with the leaf nodes attached thereto reveals that the proposed virtual infrastructure is capable of providing and ensuring an effectively stable communication service to vehicles, when there is no fixed global infrastructure.

6.3. Cooperation and performance improvement

The other objective of this structuring was to compensate for the lack of communication infrastructure and to provide all of the communication modes necessary for the continuity of cooperative applications on the road, between manned and unmanned vehicles, in non-equipped areas. In fact, the proliferation of driver assistance applications and collaborative applications that rely on communications between the vehicle and third parties (vehicle-to-everything (V2X)) will condition users in the long term. For unmanned vehicles, it even makes communications essential to ensure certain functionalities. For example, without the help provided by the cognitive abilities of a driver, the control of the trajectory and keeping in the same lane for an autonomous vehicle require cameras. Depending on the time of day (night) and the weather (rain or snow), a very precise localization (less than one meter) may also be essential. Nonetheless, the collaborative localization or ego-localization application described in Chapter 4 makes it possible for disjointed groups of vehicles to improve their localization accuracy by exchanging data on their own location as well as the associated error. In Chapter 2, we observed that CBL creates disjoint groups in an accurate manner around the branch nodes adapted to the operation of this application.

As a result of the self-organization model proposed in the case study, it becomes possible to implement by way of ad hoc communications, a solution that allows the continuity of these applications, even in degraded mode, in contexts where there is no infrastructure. However, it should be noted that several aspects have not been taken into account in the functioning of the CBL-OLSR case study protocol, particularly where effectively accounting for them would further improve its performance. As a matter of fact, the results of the comparison with QOLSR, a protocol specifically designed to provide quality of service (Sondi et al. 2013, 2014), showed that this earlier proposition actually contributes to improving link quality within the context of the *Optimized Link State Routing protocol* (OLSR) and that it is capable of better performance than CBL in terms of quality of service metrics such as throughput and end-to-end delay. Since the CBL protocol is itself implemented in OLSR, this makes it possible to envisage such an immediate integration of CBL into QOLSR to support applications incurring high-throughput and low latency requirements. This work remains one of the most interesting research avenues to this day.

However, in the presence of an existing and non-failing infrastructure, driver assistance systems will benefit in terms of performance and efficiency from using both V2I and V2V communications. In particular, a significant proportion of long-distance communications are more likely to be assured by the infrastructure. The CBL *clustering* algorithm can therefore be adapted so as to have a coordinated and optimal distribution of V2V and V2I communication tasks in routing collaborative driving automation applications according to expected traffic loads and their priority. The scientific difficulties are therefore manifold. It will be necessary to study and formalize the requirements of distributed location and perception applications, as well as the road environment, on communication metrics in the form of quality of service constraints. The study will then focus on the development of a method for characterizing traffic for the purposes of its distribution between V2I and V2V communications, as well as on the optimization of the routing protocol to meet these constraints (throughput, end-to-end delay, jitter, packet loss rate, packet error rate, packet retransmission rate, power consumption, security level, message priority, etc.). The *clustering* initially proposed in CBL will need to be modified, adapted and improved to make this distribution possible in an optimal way. The hybrid *Vehicle to Vehicle* (V2V) and *Vehicle to Infrastructure* (V2I) architecture making use of CBL (Figure 6.1) could be based on *Road Side Unit* (RSU) stations (such as 5G gNBs, for instance) deployed along the roads and which would automatically function as branch nodes. In the areas covered, the CBL structure could cease or continue to operate as an extension of the infrastructure. When stations do not exist in some areas, the CBL structure would continue to function as described in Chapter 2 of this book. Admittedly, the integration of native V2V communications into the operation of 5G today leaves a little less room for the potential standardization of concurrent solutions. Nevertheless, it may prove useful to carry on this work from the perspective of the operating frameworks set by 5G aiming at a possible integration of the solutions contemplated in the standard.

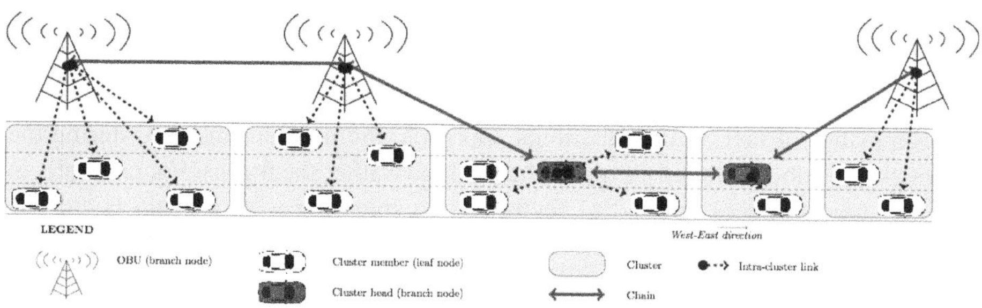

Figure 6.1. *Proposition for a hybrid V2V and V2I structure with CBL. For a color version of this figure, see www.iste.co.uk/wahl/design.zip*

6.4. Evolution of design and evaluation methods

The advent of 5G specifically carries the ambition of a very wide coverage, including outside cities, and assimilates all communication modes, especially those concerning vehicles, that is vehicle-to-infrastructure (V2I) and vehicle-to-vehicle (V2V), while capable of low latency and many other performances at levels largely adapted to vehicular communications. Combined with paradigms such as *Edge computing*, *Fog computing* and *Named Data Networking* (NDN), 5G is expected to enable faster access to devices, services, applications and data, and thus eventually establish itself as the mobile communication system for all smart city applications and consequently for communications between land transport systems.

The first consequence of this announced hegemony is to shift the focus of future research in terms of protocol design to the higher layers (transport, application, etc.) by focusing on aspects such as safety, quality of service and security, rather than at the network layer where the maturity of the solutions integrated by 5G and the expected performance suggest a stabilization over a long period of time. These expected performances make obsolete the development of virtual laboratories whose sole purpose is to compare wireless communication technologies based on performance criteria, as we have been able to present in particular in Chapters 3 and 4. The relevance of future research in terms of evaluation must imperatively include approaches for analyzing the guarantees (Fakhfakh et al. 2019) that can be provided to applications relying on communication networks, in particular in terms of quality of service, security, safety, etc.

Formal modeling precisely allows for concepts and tools adapted to these objectives at the design and evaluation level of communication protocols for land transport systems. In Chapter 5, we illustrated the application of formal methods, in this case Event-B, to the analysis of the properties of a communication protocol, namely that of the case study. In general, the formal approach proposes such a different point of view from the functional approach exposed in Chapter 2, that it has a major impact on both the design and evaluation of protocols.

As a matter of fact, in the functional approach, the mechanisms used during the design process do not directly express the goal, but rather the means. For example, in the case study protocol, instead of focusing on expressing what a chain is and the properties expected therefrom, the functional approach implements actions that organize the creation of the chain, its maintenance and repair so that it constantly corresponds to the vision that the designer has of it. Consequently, the evaluation consists, for each scenario under consideration, of executing the planned actions (the algorithm) and comparing the result obtained with the set goal. For instance, an initial evaluation can be performed using a simple graph. It then consists of observing the constructed chain by applying the planned actions in comparison with the best possible chain that could possibly be built on the same graph. This thus leads to the idea that an evaluation of this type corresponds to a performance evaluation. This is the vast majority of evaluations proposed in the works that adopt a functional approach to protocol design. It is naturally always possible to model the expected actions of conventional algorithms on graphs and to proceed to the proof of certain properties induced by these actions and which could be verified at any moment. Similarly, other analytical models can be used to model, at least on some parts of the protocol, the planned actions and to deduce thereof intrinsic properties as well as performance indexes. Simulation is one of the tools that facilitates this type of evaluation to be carried out. However, if it were necessary to determine by simulation alone that a suspected property is effective, the protocol would have to be exhaustively simulated with all conceivable scenarios. Obviously, such an approach would be all the more costly as there would be properties to be verified and possible scenarios. In this respect, this is precisely why the contribution of formal methods is decisive.

In the formal approach, the designer directly expresses the nature of the objects and the relations that link them together and can express the goal aimed at through properties on these objects and relations. It is therefore no longer necessary to continuously monitor the dynamics of the protocol, but simply to ensure that in each of the identified states the expected properties are satisfied. To this end, formal methods provide concepts and mechanisms that verify the properties expressed (so-called proof obligations) at every stage of design and establish which ones are proven and which are not (so-called residual proof obligations (RPOs)) relatively to a set of basic assumptions (axioms). Interactive proof mechanisms can be used to process OPRs and attempt to prove them based on expert-provided knowledge in order to complement basic assumptions. OPRs that cannot be proven, including by the interactive method, cannot be considered as protocol properties. It is therefore understandable that all of the properties finally proven on the formal model of the protocol are intrinsic protocol properties, provided that all conceivable scenarios verify the basic assumptions of the model. It is precisely at this level that the formal approach could prove to be flawed. How can it be established with certainty that none of the possible scenarios could possibly unveil flaws a basic assumption? More generally, how can we detect expressed properties that are not actual properties because they would have been proven from unverified assumptions in certain scenarios? Animation tools in formal tools can partially answer these questions, as we have shown in Chapter 5. Nevertheless, the difficulty of animating formal models at very large scales, which are those of applications for vehicular networks, was also mentioned therein and suggests an orientation toward coupling with simulation, as the latter is one of the tools that could have a contribution of interest. It could even, based on observables introduced into the simulation model, provide new knowledge about the protocol and the system in which it is used, in this case an ad hoc network of vehicles, and make it possible to feed an interactive proof loop instead of a human expert.

These findings open up interesting avenues. In particular, we set out to define a new approach coupling formal methods and simulation in the design and evaluation of communication protocols. As illustrated by Figure 6.2, the idea is to proceed by the following phases:

– The first phase consists of defining a set of basic assumptions that would be verified by all conceivable scenarios and achieving a formal model of the protocol on this basis.

– The second phase consists of applying the mechanisms and tools of formal methods (refinement, automatic proof, interactive proof, etc.) in order to obtain a stabilized formal model of the protocol.

– The third phase consists of constructing a simulation model equivalent to the stabilized formal model. The point is to find both the functional simulation model of the protocol, an expression of the proven properties, as well as a representation of the residual proof obligations.

– The fourth phase corresponds to the simulation of the model on all conceivable scenarios. This phase is the one that infers several discussions. First, this simulation reintroduces a cost that the formal approach should have allowed us to save; however, the number of properties to be verified is reduced only to the OPRs, since most of the properties have already been proven using the formal model. Then, by introducing observables on the elements of the model that represent the basic assumptions and properties initially proven in the formal model, the simulation can detect those that are possibly not satisfied in certain scenarios and reveal errors in the design of the formal model or in the transformation of the formal model into a simulation model. This would allow them to be corrected by returning to the appropriate step. Finally, the properties that were initially proven and that remain proven at the end of the simulation cycle are actually the established properties of the protocol. They may be eventually complemented by OPRs, the proof of which has been obtained from the additional knowledge obtained on the model by simulation.

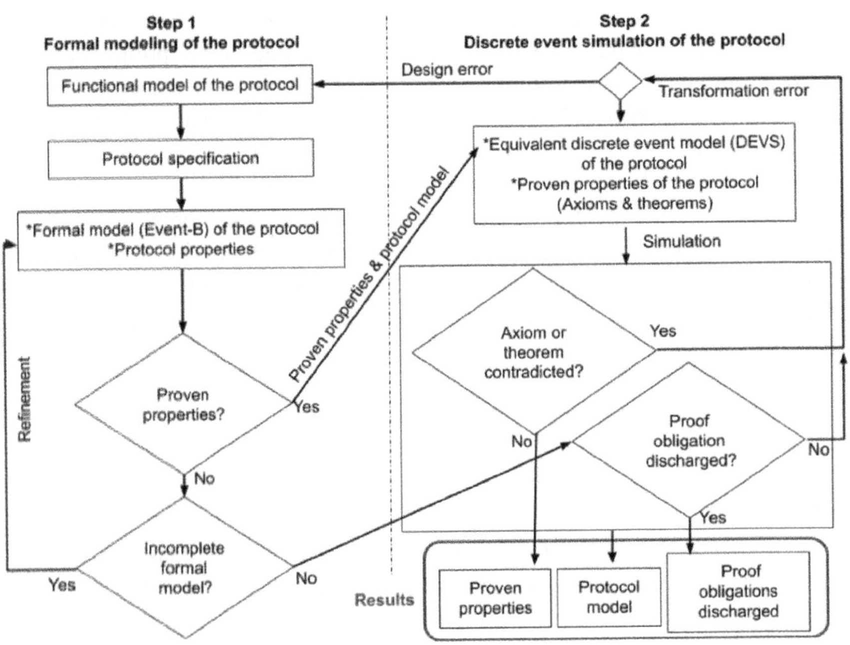

Figure 6.2. *Formal modeling/discrete event-based simulation coupling. For a color version of this figure, see www.iste.co.uk/wahl/design.zip*

This approach that we have proposed is not the first attempt at coupling formal modeling and simulation. In fact, Yacoub et al. (2016) has already proposed an approach consisting of integrating discrete-event simulation mechanisms into a formal modeling tool. The objective was to overcome the difficulty posed by the animation of large models in formal tools by inserting the simulation of a simulation model equivalent to the formal model between two formal proof steps. This approach effectively solves part of the problems associated with the animation of formal models and scalability. However, it creates issues regarding the extension of this formal/simulation coupling approach.

The DEVS (*discrete event system specification*) formalism (Zeigler et al. 2000), also used in the proposition of Yacoub et al. (2016), leads to creating a multi-modeling environment, that is, making it possible to achieve modeling from various models, each described in a different formalism. The *virtual laboratory environment* (VLE) tool proposed by Quesnel et al. (2009), for example, provides a multi-modeling environment in which the various atomic models can be realized in a specific formalism. Therefore, the formal model of the protocol could simply be integrated into multi-modeling in VLE, provided that the underlying formalism is coupled to DEVS, and subjected to interactions with models of other components that do not lend themselves to formal modeling. In the context of vehicular networks, examples can be found that include propagation models, models of particular communication technologies, mobility models, etc. More broadly speaking, through HLA (*high-level architecture*) and its RTIs (*real-time infrastructures*), the DEVS formalism also makes it possible to connect a DEVS simulation with specialized simulators for individual components (such as MATLAB for propagation models, OPNET or NS3 for communications and SUMO for mobility models). The goal is to virtually immerse the system into a realistic environment during the evaluations, as we did in Chapters 3 and 4. It therefore seems to us that in order to preserve the broadest extensibility in the formal/simulation approach, it is more appropriate to integrate the formalism underlying formal modeling into a DEVS multi-modeling approach rather than resorting to simulation from the formal model as achieved by Yacoub et al. (2016).

In addition to the formal model proposed in this chapter, we have thus proposed a first DEVS model to move toward multi-modeling. A detailed presentation of this proposition is available in Chebbi (2019), but we limit its presentation here to a summary allowing for further discussion. VLE uses DEVS modeling based on atomic models that can be coupled. As illustrated in Figure 6.3, we have introduced the following three atomic models:

– *Space* manages the placement of vehicles in space. It is possible to finely model the propagation, calculate the distances between vehicles and determine the existence and quality of the links that are notified to the *controller* model.

– *Vehicle* is an atomic model that represents a vehicle. It is possible to instantiate as many as necessary to achieve the desired density in the VANET. Each *vehicle* instance has access to the trajectory database (*Mobility Management* is not a real atomic model, but a simple schematic representation of this database) from which it can retrieve its trajectory for the entire simulation. Each time the simulation executes events related to a given vehicle, and it notifies *space* of its future spatial coordinates. The algorithms of the CBL case study protocol are implemented and executed in the *vehicle* model. Instances of *vehicle* can communicate with each other by sending messages to one another, as long as there is a link between them, which enables the HELLO and TC messages to be implemented.

– *Controller* is an atomic supermodel that can act on objects in the coupled model. It is capable of creating instances of the other models. As such, it creates a *vehicle* instance each time a new vehicle is introduced in the simulation. It also creates or removes links between instances of *vehicle* based on the information provided by *space*.

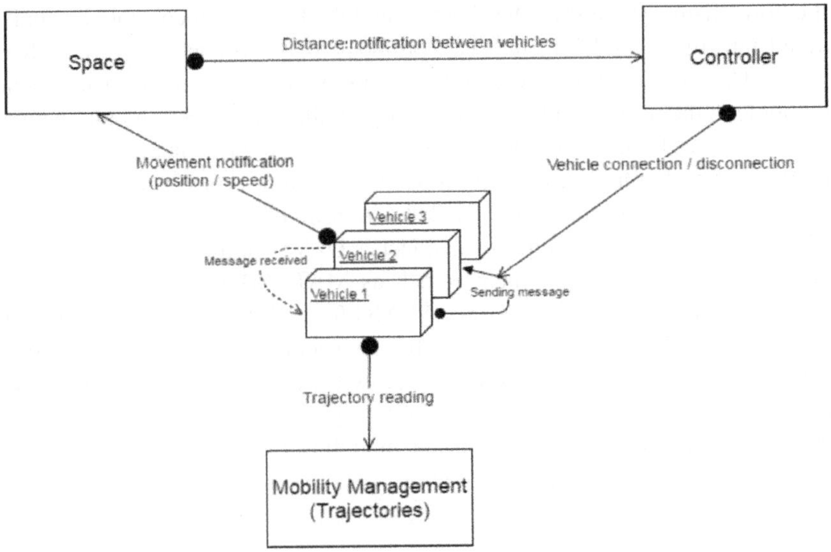

Figure 6.3. *VANET DEVS model in VLE (Chebbi 2019)*

Each atomic model makes it possible to define variables that collect various statistics throughout the simulation depending on what is happening in it. The model was validated (Chebbi et al. 2018) by comparison by simulating it on the same scenarios as those presented in Chapter 3. The *vehicle* and *controller* models group properties at the protocol level and at the global ad hoc network level, respectively. These two models would therefore be concerned by the formalism underlying formal modeling in multi-modeling.

A first problem that is not addressed, and which we intend to explore later, concerns the identification of relationships, in the formal model, between the properties described at the object level (in this case in the protocol, local scale), represented by the *vehicle* model, and those described at the system level (in this case, the ad hoc network of vehicles, global scale), represented by the *controller* model. The formal tool will actually provide the proofs of properties for each of the scales. Our objective of resorting to simulation for verifying through an exhaustive list of scenarios that the properties predicted at the global scale are all verified, given that the properties defined at the local scale are also verified, should allow us to define, in the mechanisms for transforming models from formal to simulation, relationships between properties at the two scale levels.

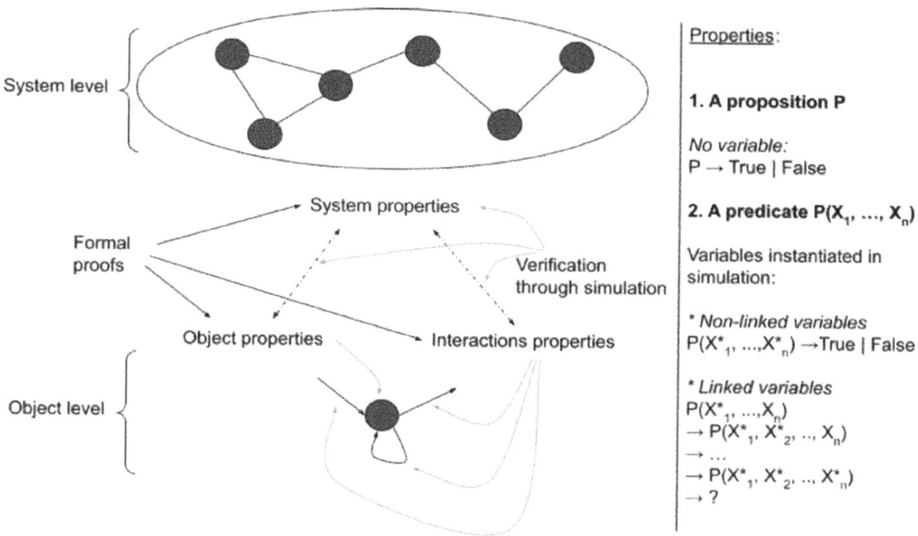

Figure 6.4. *Illustrations of the different properties. For a color version of this figure, see www.iste.co.uk/wahl/design.zip*

The second problem that remains to be investigated concerns the representation of residual proof obligations (non-automatically proven or OPR) of the formal model in a form that should be adapted to simulation. In actual fact, formal tools are capable of performing an interactive proof of OPRs by providing the missing knowledge to the automatic prover. Sometimes, a piece of data immediately triggers the proof: the question raised by the prover can thus be thought of as a one-variable predicate. Consequently, the data provided make it possible to instantiate the variable and results in the transformation of the predicate into a proposition. On the other hand, the data sometimes lead to a new question, independently of the previous one; the question can then be thought of as a predicate with several independent variables. Instantiating all of the variables simultaneously would allow the prover to provide the proof. In these first two cases, it is then possible to consider the representation of the OPR as a simple function of one or more variables in the DEVS simulation. On the other hand, if we consider a scenario in which the question posed by the prover itself depends on the first piece of data provided during the interactive proof, then the appropriate representation looks more like a tree whose traversal would depend on the data provided at each depth level. These few examples show that research in this direction remains open, but the idea behind it can be expressed simply as follows: if the model is built in such a way that certain variables express questions, then the simulation will be able to answer them by isolating the admissible values for these variables in the variety of simulated scenarios. As part of our future research related to these problems, the calculus of predicates is one of the avenues to be explored.

6.5. Conclusion

Initially developed to compensate for the lack of communication network infrastructure in areas too far away from urban centers to ensure cost-effectiveness, as well as in the event of failure of an existing infrastructure, vehicle ad hoc networks are now required to play a more permanent role in time and space. In fact, the new technologies envisioned for future networks in land transport, in particular 5G, suggest that a much wider coverage of the road network will be possible, away from urban centers, around all agricultural and industrial activity centers; in short, wherever other applications concerned by 5G will be deployed.

Therefore, on the one hand, ad hoc networks for vehicles will continuously operate in these areas in addition to the infrastructure deployed there, so as to reduce the load on the said infrastructure. Furthermore, they will provide greater responsiveness to applications that require it through so-called V2V direct communication, rather than transiting through the infrastructure. In the other hand, ad hoc networks will continue their initial mission of compensating for the lack of infrastructure. The protocols that organize them however will have to be conceived in such a way as to allow more frequently both integration with the infrastructure when vehicles reach an area covered by that infrastructure, and reorganization into an ad hoc network when a group of vehicles leave a covered area.

At the same time, this more important role is finally a real opportunity for ad hoc communications to be used on a large scale in consumer applications and to take advantage of 5G, which is expected to be far more performing than previous technologies. It also represents a shift from the previously expected requirements on routing protocols in terms of performance to requirements that will be further expressed in qualitative terms, including quality of service and safety and security. We have therefore striven to develop our design and evaluation methods toward an approach coupling formal modeling and simulation in order to be able to account for both the performance and the expected properties of communication protocols for the land transport systems of the future. The ultimate goal in mind is for communication networks to be in particular shared with the railway system where formal methods have historically already been widely used owing to the fact that this sector considers safety and security as sensitive issues.

6.6. References

Chebbi, E. (2019). De la modélisation formelle à la simulation à évènements discrets : application à la conception et à l'évaluation de protocoles sûrs et sécurisés pour les communications dans les transports. PhD Thesis, Université du Littoral Côte d'Opale, Dunkirk [Online]. Available at: http://www.theses.fr/2019DUNK0538.

Chebbi, E., Sondi, P., Ramat, E., Rivoirard, L., Wahl, M. (2018). Simulation of a clustering scheme for vehicular ad hoc networks using a DEVS-based virtual laboratory environment. *Procedia Computer Science*, 130, 344–351. doi: 10.1016/j.procs.2018.04.049.

Fakhfakh, F., Tounsi, M., Mosbah, M. (2019). An evaluative review of the formal verification for vanet protocols. In *2019 15th International Wireless Communications Mobile Computing Conference (IWCMC)*. IEEE, Tangier. doi: 10.1109/ IWCMC.2019.8766783.

Quesnel, G., Duboz, R., Ramat, É. (2009). The virtual laboratory environment – An operational framework for multi-modelling, simulation and analysis of complex dynamical systems. *Simulation Modelling Practice and Theory*, 17(4), 641–653.

Sondi, P., Gantsou, D., Lecomte, S. (2013). Design guidelines for quality of service support in optimized link state routing-based mobile ad hoc networks. *Ad Hoc Networks*, 11(1), 298–323. doi: 10.1016/j.adhoc.2012.06.001.

Sondi, P., Gantsou, D., Lecomte, S. (2014). A multiple-metric QoS-aware implementation of the optimised link state routing protocol. *International Journal of Communication Networks and Distributed Systems*, 12(4), 381–400. doi: 10.1504/IJCNDS.2014.062227.

Yacoub, A., El-Amine Hamri, M., Frydman. C. (2016). Using DEv-PROMELA for modelling and verification of software. In *Proceedings of the 2016 ACM SIGSIM Conference on Principles of Advanced Discrete Simulation (SIGSIM-PADS '16)*, 245–253. ACM, New York. doi: 10.1145/2901378.2901388.

Zeigler, B.P., Kim, T.G., Praehofer, H. (2000). *Theory of Modeling and Simulation*. Academic Press, Cambridge.

List of Authors

Patrick SONDI
CERI SN
IMT Nord Europe
Lille
and
LISIC
Université du Littoral-Côte-d'Opale
Calais
France

Martine WAHL
COSYS-LEOST
Université Gustave Eiffel
Lille
France

Index

A, C

D

E

I, L

M, N

P

R

S

T

V

Printed and bound by CPI Group (UK) Ltd, Croydon, CR0 4YY
13/09/2026
14960998-0003